202
Advances in Polymer Science

Advances in Polymer Science
Recently Published and Forthcoming Volumes

Peptide Hybrid Polymers

Volume Editors: Harm-Anton Klok · Helmut Schlaad

With contributions by

L. Ayres · T. J. Deming · J. C. M. Van Hest · K. Kataoka
H.-A. Klok · S. Lecommandoux · D. W. P. M. Löwik
K. Osada · H. Schlaad · J. M. Smeenk

 Springer

The series *Advances in Polymer Science* presents critical reviews of the present and future trends in polymer and biopolymer science including chemistry, physical chemistry, physics and material science. It is adressed to all scientists at universities and in industry who wish to keep abreast of advances in the topics covered.

As a rule, contributions are specially commissioned. The editors and publishers will, however, always be pleased to receive suggestions and supplementary information. Papers are accepted for *Advances in Polymer Science* in English.

In references *Advances in Polymer Science* is abbreviated *Adv Polym Sci* and is cited as a journal.

Springer WWW home page: springer.com
Visit the APS content at springerlink.com

Library of Congress Control Number: 2006921903

ISSN 0065-3195
ISBN-10 3-540-32567-0 Springer Berlin Heidelberg New York
ISBN-13 978-3-540-32567-3 Springer Berlin Heidelberg New York
DOI 10.1007/11677208

Springer is a part of Springer Science+Business Media

springer.com

© Springer-Verlag Berlin Heidelberg 2006
Printed in Germany

Cover design: *Design & Production* GmbH, Heidelberg
Typesetting and Production: LE-TEX Jelonek, Schmidt & Vöckler GbR, Leipzig

Printed on acid-free paper   02/3100 YL – 5 4 3 2 1 0

Advances in Polymer Science
Also Available Electronically

For all customers who have a standing order to Advances in Polymer Science, we offer the electronic version via SpringerLink free of charge. Please contact your librarian who can receive a password or free access to the full articles by registering at:

springerlink.com

If you do not have a subscription, you can still view the tables of contents of the volumes and the abstract of each article by going to the SpringerLink Homepage, clicking on "Browse by Online Libraries", then "Chemical Sciences", and finally choose Advances in Polymer Science.

You will find information about the

– Editorial Board
– Aims and Scope
– Instructions for Authors
– Sample Contribution

at springer.com using the search function.

Preface

Nature is superior to mankind, including polymer scientists, in many respects. While nature produces perfectly monodisperse and sequence-specific polypeptide and nucleotide polymers, polymer chemists are still struggling to find better methods for controlled or "living" polymerization and have yet to find strategies to control the monomer sequence beyond that of simple random and block copolymers. Natural polymers, and in particular proteins, also beat synthetic polymers in terms of structure formation. Although much progress has been made with generating complex nanoscale structures using synthetic block copolymers, the hierarchically organized tertiary and quaternary structures of proteins are still unmatched in their complexity. Natural materials also often outperform their synthetic counterparts. The catalytic activity of many enzymes and the mechanical properties of spider silk, for example, are unparalleled. Interestingly, for proteins the macroscopic properties are intimately related to chain length and monomer sequence since these control protein folding and structure formation. For synthetic polymers such exquisite (molecular) structure–property relationships have yet to be developed.

Over the past decade or so, these remarkable achievements by nature have been recognized by the polymer science community. This has led to an increased interest in the use of biological concepts to synthesize polymers or to control the structure and properties of synthetic polymers. Of particular interest are peptide hybrid polymers. Combining peptide and synthetic polymer segments into a single macromolecule offers interesting possibilities to synergize the properties of the individual components and to compatibilize bio- and synthetic systems.

This volume of *Advances in Polymer Science* is an attempt to provide an overview of the state of the art in the area of peptide hybrid polymers. The five articles in this volume cover a broad range of topics, from chemical and biological synthesis, to solution and solid-state self-assembly, to medical applications.

We would like to express our sincere thanks to all authors and reviewers who contributed to this volume for their excellent work. We hope that the articles will inspire further development in this exciting field.

Lausanne and Golm, April 2006

Harm-Anton Klok
Helmut Schlaad

Contents

Adv Polym Sci (2006) 202: 1–18
DOI 10.1007/12_080
© Springer-Verlag Berlin Heidelberg 2006
Published online: 23 February 2006

Polypeptide and Polypeptide Hybrid Copolymer Synthesis via NCA Polymerization

Timothy J. Deming

Department of Bioengineering, University of California, Los Angeles,
420 Westwood Plaza, 7523 Boelter Hall, Los Angeles, CA 90095, USA
demingt@seas.ucla.edu

Abstract This article summarizes recent developments in the synthesis of polypeptides and hybrid peptide copolymers. Traditional methods used to polymerize α-amino acid-N-carboxyanhydrides (NCAs) are described, and limitations in the utility of these systems for the preparation of polypeptides are discussed. Recently developed initiators and methods are also discussed that allow polypeptide synthesis with good control over chain length, chain length distribution, and chain-end functionality. The latter feature is particularly useful for the preparation of polypeptide hybrid copolymers. The methods and strategies for the preparation of such hybrid copolymers are described, as well as analysis of the synthetic scope of the different methods. Finally, issues relating to obtaining these highly functional copolymers in pure form are detailed.

Keywords Polypeptide · Block copolymer · *N*-carboxyanhydride · Living polymerization · Hybrid copolymer

Abbreviations

NCA	α-amino acid N-carboxyanhydride
AM	activated monomer
GPC	gel permeation chromatography
NACE	non-aqueous capillary electrophoresis
PBLG	poly(γ-benzyl-L-glutamate)
PMLG	poly(γ-methyl-L-glutamate)

PMDG   poly(γ-methyl-D-glutamate)
PML/DG   poly(γ-methyl-*rac*-glutamate)
PZLL   poly(ε-carbobenzyloxy-L-lysine)
PBLA   poly(β-benzyl-L-aspartate)
PMA   polymethylacrylate
PEG   polyethylene glycol
depe   bis(diethylphosphino)ethane

1
Introduction

Biological systems produce proteins that possess the ability to self-assemble into complex, yet highly ordered structures [1]. These remarkable materials are polypeptide copolymers that derive their properties from precisely controlled sequences and compositions of their constituent amino acid monomers. There has been recent interest in developing synthetic routes for preparation of these natural polymers as well as de novo designed polypeptide sequences to make products for applications in biotechnology (artificial tissues, implants), biomineralization (resilient, lightweight, ordered inorganic composites), and analysis (biosensors, medical diagnostics) [2, 3].

To be successful in these applications, it is important that materials can self-assemble into precisely defined structures. Peptide polymers have many advantages over conventional synthetic polymers since they are able to hierarchically assemble into stable ordered conformations [4]. Depending on the amino acid side chain substituents, polypeptides are able to adopt a multitude of conformationally stable regular secondary structures (helices, sheets, turns), tertiary structures (e.g. the β-strand-helix-β-strand unit found in β-barrels), and quaternary assemblies (e.g. collagen microfibrils) [4]. The synthesis of polypeptides that can assemble into non-natural structures is an attractive challenge for polymer chemists.

Synthetic peptide-based polymers are not new materials: homopolymers of polypeptides have been available for many decades and have only seen limited use as structural materials [5, 6]. However, new methods in chemical synthesis have made possible the preparation of increasingly complex polypeptide sequences of controlled molecular weight that display properties far superior to ill-defined homopolypeptides [7]. Furthermore, hybrid copolymers, that combine polypeptide and conventional synthetic polymers, have been prepared and combine the functionality and structure of peptides with the processability and economy of polymers [8, 9]. These polymers are well suited for applications where polymer assembly and functional domains need to be at length scales ranging from nanometers to microns. These block copolymers are homogeneous on a macroscopic scale, but dissimilarity between the block segments typically results in microphase heterogeneity yielding materials useful as surfactants, micelles, membranes, and elastomers [10]. Synthesis of simple hydrophilic/hydrophobic hybrid diblock copolymers, when dispersed in water, allows formation of peptide-based micelles and vesicles potentially useful in drug and gene delivery applications [11]. The regular secondary structures obtainable with the polypeptide blocks provide opportunities for hierarchical self-assembly unobtainable with typical block copolymers or small-molecule surfactants.

Upon examining the different methods for polypeptide synthesis, the limitations of these techniques for preparation of hybrid copolymers readily become apparent. Conventional solid-phase peptide synthesis is neither useful nor practical for direct preparation of large polypeptides (> 100 residues) due to unavoidable deletions and truncations that result from incomplete deprotection and coupling steps. The most economical and expedient process for synthesis of long polypeptide chains is the polymerization of α-amino acid-N-carboxyanhydrides (NCAs) (Scheme 1) [12, 13]. This method involves the simplest reagents and high molecular weight polymers can be prepared in both good yield and large quantity with no detectable racemization at the chiral centers. The considerable variety of NCAs that have been synthesized (> 200) allows exceptional diversity in the types of polypeptides that can be prepared [12, 13].

Since the late 1940s, NCA polymerizations have been the most common technique used for large-scale preparation of high molecular weight polypeptides [14]. However, these materials have primarily been homopolymers, random copolymers, or graft copolymers that lack the sequence specificity and monodispersity of natural proteins. The level of control in NCA polymerizations has not been able to rival that attained in other synthetic polymerizations (e.g. vinyl addition polymerizations) where sophisticated polymer architectures have been prepared (e.g. stereospecific polymers and block copolymers) [15]. Attempts to prepare block copolypeptides and hybrid block copolymers using NCAs have always resulted in polymers whose compositions did not match monomer feed compositions and that contained significant homopolymer contaminants [16, 17]. Block copolymers could only be obtained in pure form by extensive fractionation steps, which significantly lowered the yield and efficiency of this method. The limitation of NCA polymerizations has been the presence of side reactions (chain termination and chain transfer) that restrict control over molecular weight, give broad mo-

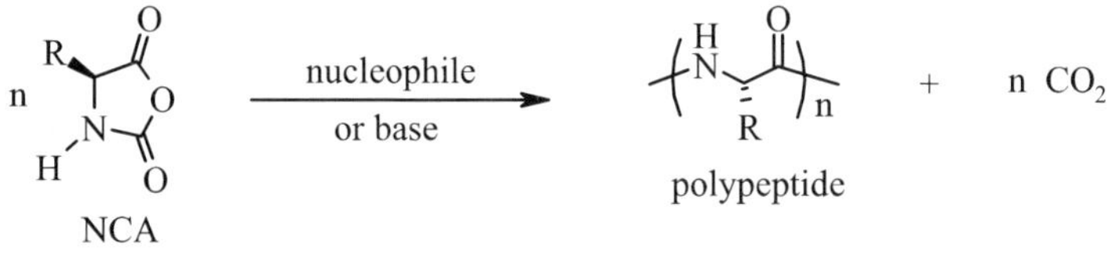

Scheme 1

lecular weight distributions, and prohibit formation of well-defined block copolymers [19, 20]. Recent progress in elimination of these side reactions has been a major breakthrough for the polypeptide materials field.

2
Polypeptide Synthesis using NCAs

2.1
Conventional Methods

NCA polymerizations are traditionally initiated using many different nucleophiles and bases, the most common being primary amines and alkoxide anions [12, 13]. Primary amines, being more nucleophilic than basic, are good general initiators for polymerization of NCA monomers. Tertiary amines, alkoxides, and other initiators that are more basic than nucleophilic, have found use since they are in some cases able to prepare polymers of very high molecular weight where primary amine initiators cannot. Optimal polymerization conditions have often been determined empirically for each NCA and thus there have been no universal initiators or conditions by which to prepare high polymers from any monomer. This is in part due to the different properties of individual NCAs and their polymers (e.g. solubility) but is also strongly related to the side reactions that occur during polymerization.

The most likely pathways of NCA polymerization are the so-called "amine" and the "activated monomer" (AM) mechanisms [12, 13]. The amine mechanism is a nucleophilic ring opening chain growth process where the polymer could grow linearly with monomer conversion if side reactions were absent (Scheme 2). On the other hand, the AM mechanism is initiated by deprotonation of an NCA, which then becomes the nucleophile that initiates chain growth (Scheme 3). It is important to note that a given system can switch back and forth between the amine and AM mechanisms many times during a polymerization: a propagation step for one mechanism is a side reaction for the other, and vice versa. It is because of these side reactions that block copolypeptides and hybrid block copolymers prepared from NCAs using

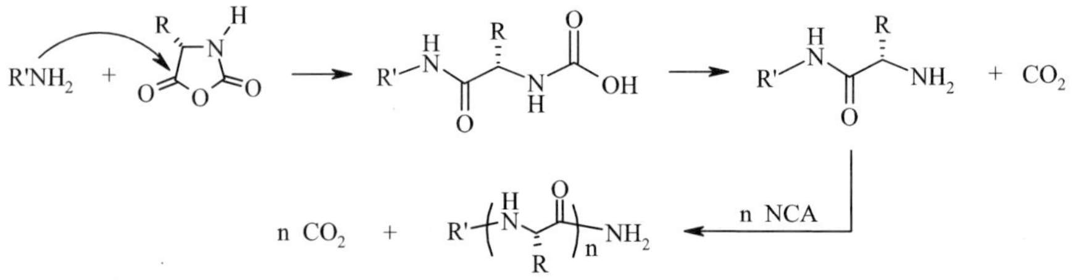

Scheme 2

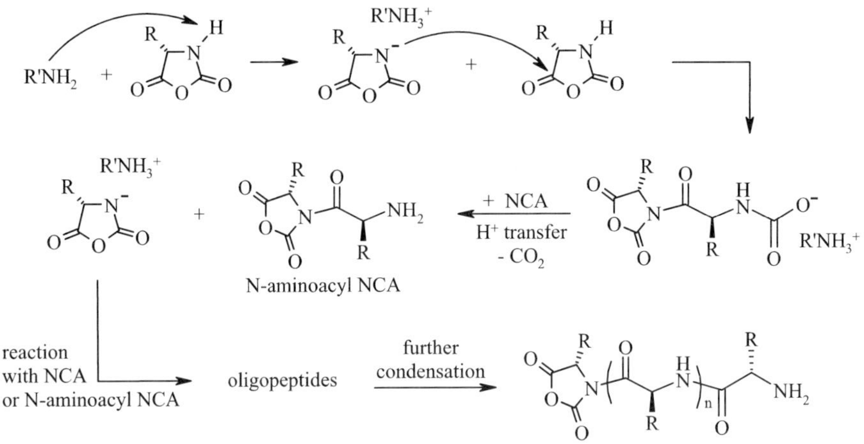

Scheme 3

amine initiators have structures different than predicted by monomer feed compositions and most likely have considerable homopolymer contamination. These side reactions also prevent control of chain-end functionality, which is crucial for preparation of hybrid copolymers.

One inherent problem in conventional NCA polymerizations is that there is no control over the reactivity of the growing polymer chain-end during the course of the polymerization. Once an initiator reacts with a NCA monomer, it is no longer active in the polymerization and the resulting primary amine, carbamate, or NCA anion end-group is free to undergo a variety of undesired side reactions. Another problem is one of purity. Although most NCAs are crystalline compounds, they typically contain minute traces of acid, acid chlorides, or isocyanates that can quench propagating chains. The presence of other adventitious impurities, such as water, can cause problems by acting as chain-transfer agents or even as catalysts for side-reactions. Overall, the sheer abundance of potential reactions present in reaction media make it difficult to achieve a living polymerization system where only chain propagation occurs.

2.2
Transition Metal Initiators

One strategy to eliminate side-reactions in NCA polymerizations is the use of transition metal complexes as active species to control addition of NCA monomers to polymer chain-ends. The use of transition metals to control reactivity has been proven in organic and polymer synthesis as a means to increase both reaction selectivity and efficiency [21]. Using this approach, substantial advances in controlled NCA polymerization have been realized in recent years. Highly effective zerovalent nickel and cobalt initiators (i.e. bpyNi(COD) [22, 23] and (PMe$_3$)$_4$Co [24]) were developed by Deming that allow the living polymerization of NCAs into high molecular weight polypeptides via an unprecedented activation of the NCAs into covalent propagating species. The metal ions can be conveniently removed from the polymers by simple precipitation or dialysis of the samples after polymerization.

Mechanistic studies on the initiation process showed that both these metals react identically with NCA monomers to form metallacyclic complexes by oxidative addition across the anhydride bonds of NCAs [22–24]. These oxidative-addition reactions were followed by addition of a second NCA monomer to yield complexes identified as six-membered amido-alkyl metallacycles (Scheme 4). These intermediates were found to further contract to five-membered amido-amidate metallacycles upon reaction with additional NCA monomers. This ring contraction is thought to occur via migration of an amide proton to the metal-bound carbon, which liberates the chain-end from the metal (Scheme 5) [25]. The resulting amido-amidate complexes were thus proposed as the active polymerization intermediates. Propagation through the amido-amidate metallacycle was envisioned to occur by initial attack of the nucleophilic amido group on the electrophilic C_5 carbonyl of an NCA monomer (Scheme 6). This reaction would result in a large metallacycle that

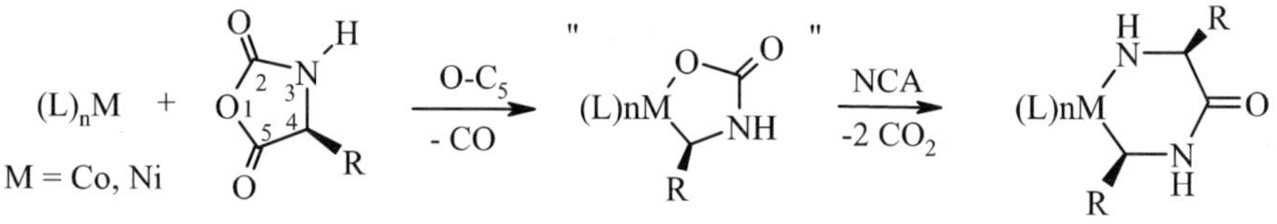

Scheme 4

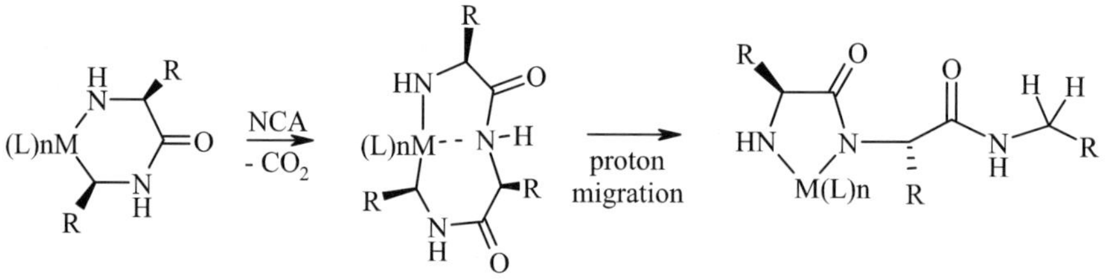

Scheme 5

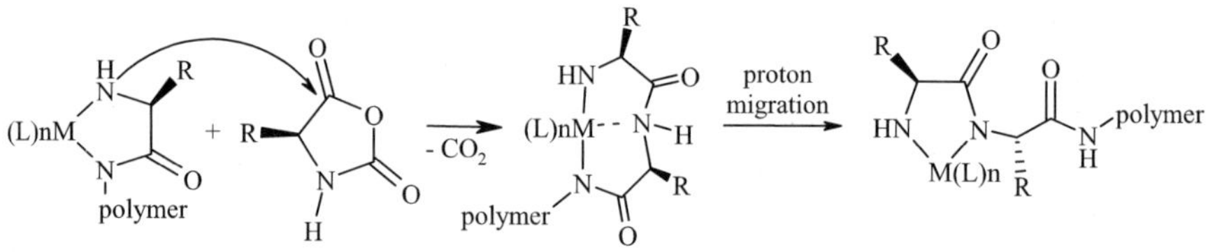

Scheme 6

could contract by elimination of CO_2. Proton transfer from the free amide to the tethered amidate group would further contract the ring to give the amido-amidate propagating species, while in turn liberating the end of the polymer chain.

In this manner, the metal is able to migrate along the growing polymer chain, while being held by a robust chelate at the active end. The formation of these chelating metallacyclic intermediates appears to be a general requirement for obtaining living NCA polymerizations using transition metal initiators. These cobalt and nickel complexes are able to produce polypeptides with narrow chain length distributions ($M_w/M_n < 1.20$) and controlled molecular weights ($500 < M_n < 500\,000$) [26]. This polymerization system is very general, and gives controlled polymerization of a wide range of NCA monomers as pure enantiomers (D or L configuration) or as racemic mixtures. By addition of different NCA monomers, the preparation of block copolypeptides of defined sequence and composition is feasible [7, 27].

2.3
New Developments

In the past two years, a number of new approaches have been reported for obtaining controlled NCA polymerizations. These approaches share a common theme in that they are all improvements on the use of classical primary amine polymerization initiators. This approach is attractive since primary amines are readily available and since the initiator does not need to be removed from the reaction after polymerization. In fact, if the polymerization proceeds without any chain breaking reactions, the amine initiator becomes the C-terminal polypeptide end-group. In this manner, there is potential to form chain-end-functionalized polypeptides or even hybrid block copolymers if the amine is a macroinitiator. The challenge in this approach is to overcome the numerous side-reactions of these systems without the luxury of a large number of experimental parameters to adjust.

In 2004, the group of Hadjichristidis reported the primary amine-initiated polymerization of NCAs under high vacuum conditions [28]. The strategy here was to determine if a reduced level of impurities in the reaction mixture would lead to fewer polymerization side reactions. Unlike the vinyl monomers usually polymerized under high vacuum conditions, NCAs cannot be purified by distillation. Consequently, it is doubtful the NCAs themselves can be obtained in higher purity under high vacuum recrystallization than by recrystallization under a rigorous inert atmosphere. However, the high vacuum method does allow for better purification of polymerization solvents and the n-hexylamine initiator. It was found that polymerizations of γ-benzyl-L-glutamate NCA (Bn-Glu NCA) and ε-carbobenzyloxy-L-lysine NCA (Z-Lys NCA) under high vacuum in DMF solvent displayed all the characteristics of a living polymerization system [28]. Polypeptides could be prepared with control over chain length, chain length distributions were narrow, and block copolypeptides were prepared.

The authors concluded that the side-reactions normally observed in amine-initiated NCA polymerizations are simply a consequence of impurities. Since the main side reactions in these polymerizations do not involve reaction with adventitious impurities such as water, but instead reactions with monomer, solvent, or polymer (i.e. termination by reaction of the amine-end with an ester side-chain, attack of DMF by the amine-end, or chain transfer to monomer) [12, 13], this conclusion does not seem to make sense. It is likely that the role of impurities (e.g. water) in these polymerizations is very complex. A possible explanation for the polymerization control observed under high vacuum is that the impurities act to catalyze side reactions with monomer, polymer, or solvent. In this scenario, it is reasonable to speculate that polar species such as water can bind to monomers or the propagating chain end and thus influence their reactivity.

Further insights into amine-initiated NCA polymerizations were also reported in 2004 by the group of Giani and coworkers [29]. This group studied the polymerization of ε-trifluoroacetyl-L-lysine NCA (TFA-Lys NCA) in DMF using n-hexylamine initiator as a function of temperature. Contrary to the high vacuum work, the solvent and initiator were purified using conventional methods and the polymerizations were conducted under nitrogen on a Schlenk line. After complete consumption of NCA monomer, the crude polymerization mixtures were analyzed by GPC and non-aqueous capillary electrophoresis (NACE). A unique feature of this work was the use of NACE to separate and quantify the amount of polymers with different chain-ends, which corresponded to living chains (amine end-groups) and "dead" chains (carboxylate and formyl end-groups from reaction with NCA anions and DMF solvent, respectively, Schemes 7 and 8). Not surprisingly, at 20 °C, the polymer products consisted of 78% dead chains, and only 22% living chains, which illustrates the abundance of side reactions in these polymerizations under normal conditions.

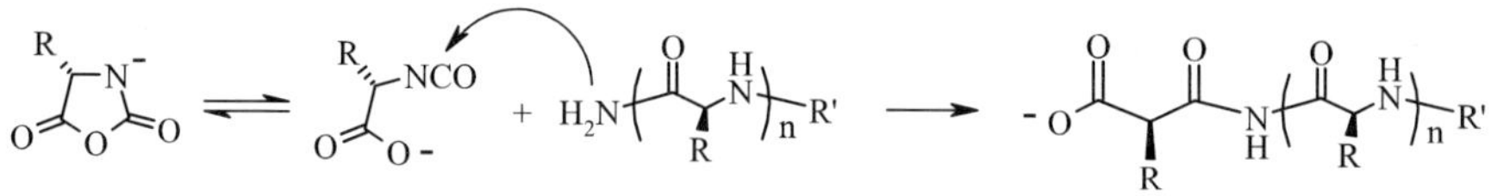

Scheme 7

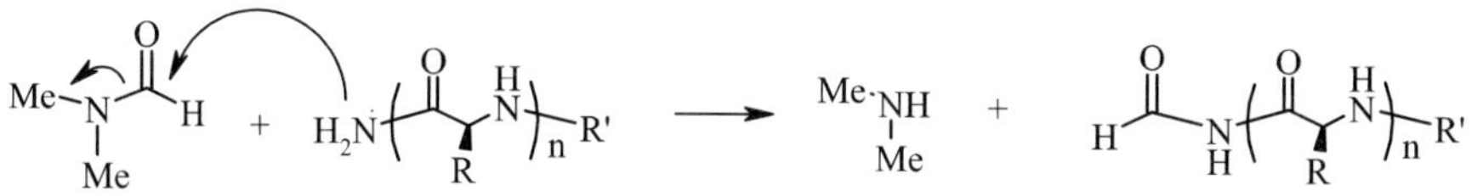

Scheme 8

An intriguing result was found for polymerizations conducted at 0 °C where 99% of the chains had living amine chain ends, and only 1% were found to be dead chains. To verify that these were truly living polymerizations, additional NCA monomer was added to these chains at 0 °C resulting in increased molecular weight and no increase of the amount of dead chains. While this was only a preliminary study and further studies need to be conducted to explore the scope of this method, this work clearly shows that the common NCA polymerization side reactions can also be eliminated by lowering temperature. The effect of temperature is not unusual, as similar trends can be found in cationic and anionic vinyl polymerizations [30]. At elevated temperature, the side reactions have activation barriers similar to chain propagation. When the temperature is lowered, it appears that the activation barrier for chain propagation becomes lower than that of the side reactions and chain propagation dominates kinetically. A remarkable feature of this system is that the elevated levels of impurities, as compared to the high vacuum method, do not seem to cause side reactions at low temperature. This result further substantiates the idea that the growing chains do not react with the adventitious impurities, but that they mainly affect these polymerizations by altering the rates of discrete reaction steps.

Another innovative approach to controlling amine-initiated NCA polymerizations was reported in 2003 by Schlaad and coworkers [31]. Their strategy was to avoid formation of NCA anions, which cause significant chain termination after rearranging to isocyanocarboxylates [12, 13], through use of primary amine hydrochloride salts as initiators. The reactivity of amine hydrochlorides with NCAs was first explored by the group of Knobler, who found that they can react with NCAs to give single NCA addition products [32, 33]. Use of the hydrochloride salt takes advantage of its diminished reactivity as a nucleophile compared to the parent amine, which effectively halts the reaction after a single NCA insertion by formation of an inert amine hydrochloride in the product. The reactivity of the hydrochloride presumably arises from formation of a small amount of free amine by reversible dissociation of HCl (Scheme 9). This equilibrium, which lies heavily toward the dormant amine hydrochloride species, allows for only a very short lifetime of reactive amine species. Consequently, as soon as a free amine reacts with

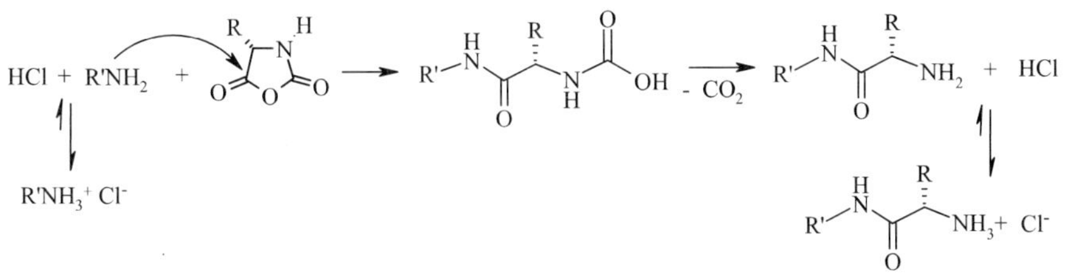

Scheme 9

an NCA, the resulting amine end-group on the product is immediately proto-nated and is prevented from further reaction. The acidic conditions also assist elimination of CO_2 from the reactive intermediate, and more importantly, suppress formation of unwanted NCA anions.

To obtain controlled polymerization, and not just single NCA addition re-actions, Schlaad's group increased the reaction temperature (40 to 80 °C), which was known from Knobler's work to increase the equilibrium con-centration of free amine, as well as increase the exchange rate between amine and amine hydrochloride [32, 33]. Using primary amine hydrochlo-ride end-capped polystyrene macroinitiators to polymerize Z-Lys NCA in DMF, Schlaad's group obtained polypeptide hybrid copolymers in 70 to 80% yield after 3 days at elevated temperature. Although these polymerizations are slow compared to amine-initiated polymerizations, the resulting polypep-tide segments were well defined with very narrow chain length distributions ($M_w/M_n < 1.03$). These distributions were much narrower than those ob-tained using the free amine macroinitiator, which argues for diminished side reactions in the polypeptide synthesis. The molecular weights of the resulting polypeptide segments were found to be ca. 20 to 30% higher than would be expected from the monomer to initiator ratios. This result was attributed to termination of some fraction of initiator species by traces of impurities in the NCA monomers, although the presence of unreacted polystyrene chains was not reported.

Although more studies need to be performed to study the scope and gen-erality of this system, the use of amine hydrochloride salts as initiators for controlled NCA polymerizations shows tremendous promise. The concept of fast, reversible deactivation of a reactive species to obtain controlled poly-merization is a proven concept in polymer chemistry, and this system can be compared to the persistent radical effect employed in all controlled rad-ical polymerization strategies [34]. Like those systems, the success of this method requires a carefully controlled matching of the polymer chain propa-gation rate constant, the amine/amine hydrochloride equilibrium constant, and the forward and reverse exchange rate constants between amine and amine hydrochloride salt. This means it is likely that reaction conditions (e.g. temperature, halide counterion, solvent) will need to be optimized to obtain controlled polymerization for each different NCA monomer, as is the case for most vinyl monomers in controlled radical polymerizations. Within these constraints, it is possible that controlled NCA polymerizations utilizing sim-ple amine hydrochloride initiators can be obtained.

3
Polypeptide Hybrid Copolymer Synthesis

3.1
Conventional Methods

The synthesis of polypeptide hybrid block copolymers was first reported by the groups of Gallot and coworkers [35] and Yamashita and coworkers [36] in the mid 1970s. These initial studies were followed by numerous reports continuing to the present on the preparation of both AB- and ABA-type block copolymers (A = polypeptide, B = synthetic domain) [8, 9]. The vast majority of these copolymers were prepared using a two-step process where a synthetic macroinitiator was first prepared with either one or both chain-ends functionalized with primary amine groups. These macroinitiators were then used to synthesize the second polypeptide domain using conventional methods for NCA polymerization. The block copolymerization is usually conducted in two stages since the synthetic chemistry of the different segments are typically incompatible.

The advantages of this approach are that most synthetic polymer segments can be prepared (typically by anionic polymerization) with controlled chain lengths, low polydispersity, and a high degree of amine functionalization on the chain-ends. Furthermore, a single batch of amine-functionalized polymer can be used to prepare a variety of different block copolymers by reaction with different NCAs. The major drawback of this method is the abundance of side-reactions in the NCA polymerization step, which leads to a high degree of polydispersity in the polypeptide segments and the formation of homopolymer contaminants. The resulting crude hybrid block copolymers must be extracted with selective solvents to remove homopolymers, and then fractionated by selective precipitation to obtain copolymers with lower polydispersity in the polypeptide domain [8, 9].

Despite the drawbacks of this method, it has been used to prepare a tremendous number of polypeptide hybrid block copolymers (Table 1), and when carefully executed provides reasonably well-defined samples. Synthetic polymer domains have been prepared by addition polymerization of conventional vinyl monomers, such as styrene and butadiene, as well as by ring-opening polymerization in the cases of ethylene oxide and ε-caprolactone. The generality of this approach allows NCA polymerization off of virtually any primary amine functionality, which was exploited in the preparation of star block copolymers by polymerization of sarcosine NCA from an amine-terminated trimethyleneimine dendritic core [37]. In most examples, the polypeptide domain was based on derivatives of either lysine or glutamate, since these form α-helical polypeptides with good solubility characteristics. These residues are also desirable since, when deprotected, they give polypep-

Table 1 Examples of polypeptide hybrid block copolymers prepared from macromolecular amine initiators [8, 9]. Block architectures are primarily AB diblocks or ABA triblocks, where the polypeptide segment is the A domain, and the macroinitiator is the B domain

Amine macroinitiator	Polypeptide segments (*Architecture*)
Polystyrene	PBLG(*AB*); PZLL(*AB*), PMDG(*ABA*)
Polybutadiene	PZLL(*AB,ABA*); PBLG(*AB,ABA*); PBL/DG (*ABA*); PML/DG (*ABA*)
Polyisoprene	PBLG(*ABA*)
Polydimethylsiloxane	PBLG(*AB, ABA*); Poly(L/D-Phe)(*AB*)
Polyethylene glycol	PZLL(*AB,ABA*); PBLG(*AB,ABA*); Poly(L-Pro)(*ABA*); PBLA(*AB,ABA*)
Polypropylene oxide	PBLG(*ABA*)
Poly(2-methyloxazoline)	PBLG(*AB*); Poly(L-Phe)(*AB*) [56]
Poly(2-phenyloxazoline)	PBLG(*AB*); Poly(L-Phe)(*AB*) [56]
Polymethyl methacrylate	PZLL(*AB*); PBLG(*AB*); PMLG(*AB*)
Polymethyl acrylate	PBLG(*AB*)
Polyoctenamer	PBLG(*ABA*)
Polyethylene	PBLG(*ABA*)
Polyferrocenylsilane	PBLG(*AB*) [57]
Poly(9,9-dihexylfluorene)	PBLG(*ABA*) [58]
Poly(ε-caprolactone)	PBLG(*ABA*); Poly(L-Phe)(*ABA*); Poly(Gly)(*ABA*); Poly(L-Ala)(*ABA*) [59]
Trimethyleneimine dendrimer	Poly(Sarcosine)(*Star dendrimer*)

tide segments that are both water soluble and conformationally responsive to stimuli such as pH and temperature [5, 6].

A few other methods have been used to prepare polypeptide hybrid copolymers. Inoue polymerized Bn-Glu NCA off of amine-functionalized styrene derivatives, and then copolymerized these end-functionalized polypeptides with either styrene or methyl methacrylate using free radical initiators to yield hybrid comb architecture copolymers [38]. Although unreacted polypeptide was identified and removed by fractionation, copolymers were obtained with polypeptide content that increased with feed ratio. There was no mention if the polypeptide interfered with the radical chemistry. In similar work, Imanishi and coworkers converted the amine-ends of polypeptides to haloacetyl groups that were used to initiate the free radical polymerization of either styrene or methylmethacrylate to yield hybrid block copolymers [39]. Studies using GPC showed that the crude product contained mixtures of copolymers and homopolymers, and so removal of the homopolymers by extraction was necessary.

In an approach combining step growth polymerization with ring opening polymerization, Uchida and coworkers prepared a linear polyurethane capped with isocyanate groups on both ends [40]. This macromonomer was then mixed with NCA and hydrazine initiator, which was designed to yield a polypeptide in situ that is capped with primary amine on both ends of the chain. This difunctional polypeptide was then meant to condense with the polyurethane to yield an $(AB)_n$ multiblock hybrid copolymer. Although the resulting materials were not characterized in detail, the presence of oligomeric contaminants was noted. Overall, it can be seen that a wide variety of polypeptide hybrid block copolymers, with a range of architectures, can potentially be obtained using conventional chemistry. However, as with any synthesis, removal of side reactions, especially from the polypeptide synthesis step, would be expected to greatly streamline the preparation of these materials and allow more precise control over segment lengths and copolymer compositions. Such fine control is especially useful, and in many cases required, for self-assembly of these copolymers into ordered nanostructures [41].

3.2
Controlled Polymerizations

The recent improvements in amine-initiated NCA polymerizations utilizing high vacuum and low temperature have, to date, not been used to prepare polypeptide hybrid block copolymers. However, the amine hydrochloride initiators developed by Schlaad and coworkers (vide infra) were in fact macroinitiators and gave well-defined polystyrene-poly(ε-carbobenzyloxy-L-lysine), PS-PZLL, block copolymers in good yield [31]. This chemistry has been extended by Lutz and coworkers who polymerized Bn-Glu NCA and β-benzyl-L-aspartate NCA (Bn-Asp NCA) using amine hydrochloride-functionalized PEG [42]. Although the chain length distributions of the resulting copolymers were found to be very narrow ($M_w/M_n < 1.05$), the polymerizations were noted to be sluggish, with oligopeptide chain lengths found to be much less than targeted (10 residues) after three days at 40 °C. Small amounts of peptide oligomers were also present in the samples that had to be removed by selective precipitation.

The transition metal initiators for NCA polymerization described above should provide a means for controlled synthesis of polypeptide hybrid block copolymers. However, a limitation of this methodology when using zerovalent metal complexes as initiators is that the active propagating species are generated in situ, where the C-terminal end of the polypeptide is derived from the first NCA monomer. Consequently, this method does not allow attachment of a synthetic polymer (e.g. amine functionalized) to the carboxyl chain-end. For this reason, Deming and coworkers pursued alternative methods for direct synthesis of the amido-amidate metallacycle propagating species and developed allyloxycarbonylaminoamides as universal precursors to amido-amidate nickelacycles. These simple amino acid derivatives undergo tandem oxidative-additions to nickel(0) to give active NCA polymerization initiators (Scheme 10) [43]. These complexes were found to ini-

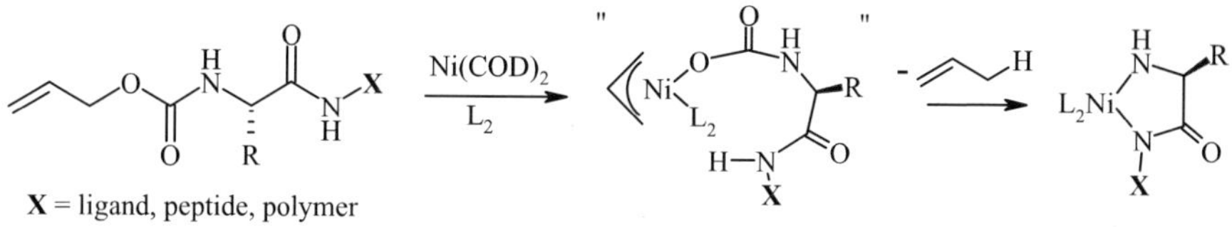

X = ligand, peptide, polymer

Scheme 10

tiate polymerization of NCAs yielding polypeptides with defined molecular weights, narrow molecular weight distributions, and with quantitative incorporation of the initiating ligand as a C-terminal end-group. This chemistry provides a very facile means to incorporate diverse molecules such as peptides, oligosaccharides or other ligands onto the chain-ends of polypeptides via a robust amide linkage, and was further elaborated by Menzel's group to grow polypeptides off of polystyrene particles [44].

This chemistry is also well-suited for the synthesis of polypeptide hybrid block copolymers by growth of polypeptides from amino-functionalized synthetic polymers. This approach is very similar to that used previously for preparation of these hybrid materials, except that the nickel-based initiators provide additional control over the polypeptide synthesis. This chemistry was used to prepare a variety of ABA-type triblock hybrid copolymers from $\alpha\omega$-diamine-functionalized polymers, where A was poly(γ-benzyl-L-glutamate), PBLG, and B included each of the following: polyoctenamer, polyethylene, PEG, and PDMS [45]. Use of this chemistry gave copolymers where the polypeptide segment lengths could be varied with good control. Furthermore, no unreacted homopolymers or homopolypeptides could be detected. This methodology for preparation of block copolymers appears to be general and can be used with a wide range of amino-terminated polymers (Scheme 11). For example, Deming recently reported the preparation of poly(methylacrylate)-PBLG, PMA-PBLG, diblock copolymers that were prepared using amine end-functionalized PMA obtained using controlled radical polymerization [46]. The two-step conversion of the amine group into the nickelacycle initiator allowed controlled addition of Bn-Glu NCA to the PMA chains with narrow overall chain length distributions ($M_w/M_n < 1.2$).

As an extension of this work, Deming also developed a means to end-cap living polypeptide chains with electrophilic reagents. When a macromolecular electrophile is used, the resulting product is a polypeptide hybrid block

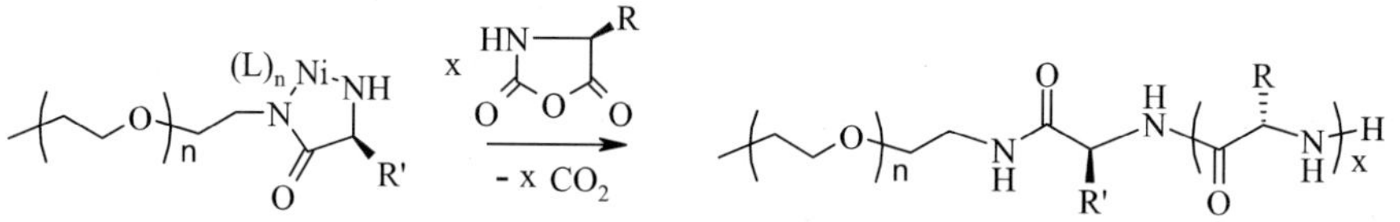

Scheme 11

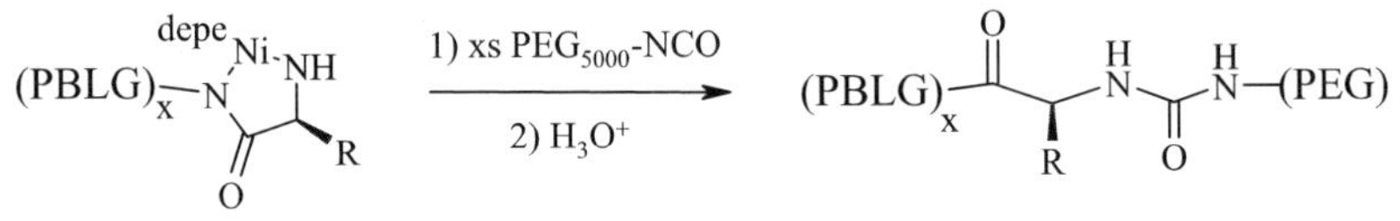

Scheme 12

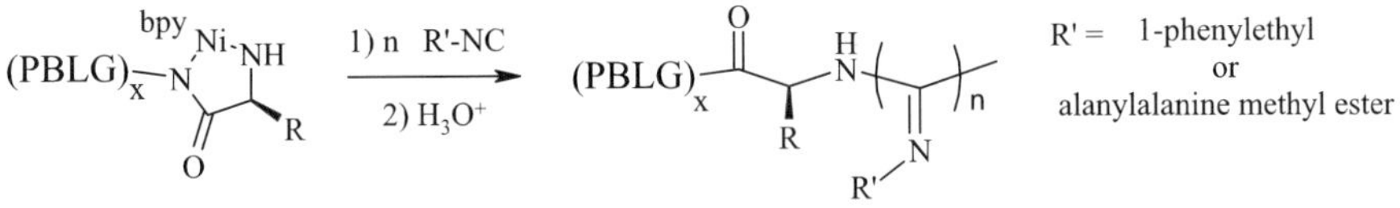

Scheme 13

copolymer. It is well known in NCA polymerizations that electrophiles, such as isocyanates, act as chain-terminating agents by reaction with the propagating amine chain-ends [12, 13]. Deming and coworkers reported that the reactive living nickelacycle polypeptide chain-ends could be quantitatively capped by reaction with excess isocyanate, isothiocyanate, or acid chloride [47]. Using this chemistry, they prepared isocyanate end-capped PEG and reacted this, in excess, with living PBLG to obtain PBLG-PEG diblock copolymers (Scheme 12). Reaction with living ABA triblock copolymers (vide infra) gave the corresponding PEG-capped CABAC hybrid pentablock copolymers, where A was PBLG, B was polyoctenemer, PEG or PDMS, and C was PEG. Since excess PEG was used to end-cap the living polypeptide chains, the pentablock copolymers required purification, which was achieved by repeated precipitation from THF into methanol. Recently, Cornelissen and coworkers reported the polymerization of isocyanides from living nickelacycle end-capped PBLG (Scheme 13) [48]. This synthesis is remarkable in that the nickel species were used for stepwise polymerization of both NCAs and isocyanides in a one-pot procedure. Overall, it can be seen that the use of controlled NCA polymerization allows formation of very complex hybrid block copolymer architectures that rival those prepared using any polymerization system.

4
Hybrid Copolymer Deprotection and Purification

Although quite complex hybrid block copolymer architectures can now be synthesized, obtaining these materials in a state of high purity typically requires additional measures. As discussed above, many of the hybrid copolymers contain homopolymer impurities, which must be removed by selective solvent extractions or fractional precipitation when possible. Since conventional NCA polymerizations also usually give polypeptide segments with large chain length distributions, these samples are ideally also fractionated to give samples of well-defined composition. An additional purification issue arises from the amphiphilic nature of many of these copolymers, e.g. PEG-PBLG. Such polymers tend to associate in most solvents leading to trapped solvents or solutes in the copolymer sample, which can complicate analytical studies. In the case of transition metal-initiated polymerizations, removal of the metal from the sample is also important for most applications. For rigorous purification of these amphiphilic copolymers, the author has found exhaustive dialysis of the samples against deionized water to be very effective at removing small molecule contaminants. In cases where a polymer segment can bind strongly to metals such as Co^{2+} and Ni^{2+}, the addition of a potent metal chelator, such as EDTA, in the early stages of dialysis was found to be sufficient to remove all traces of the metal ions.

A highly useful feature of the polypeptide segments in these hybrid copolymers is their functionality. The common naturally occurring amino acids contain numerous acidic and basic functional groups that provide interesting pH responsive character to these materials. These functional groups are masked by protecting groups before synthesis of the NCA monomers, since they will typically interfere with polypeptide synthesis or NCA stability [12, 13]. Consequently, these protecting groups must be removed after polymerization if one is to utilize the functional group chemistry. The first concern with polypeptide deprotection is whether or not all the protecting groups have been removed. Small amounts of residual protecting groups can significantly influence the resulting polypeptide properties, especially since the protecting groups are typically hydrophobic and the deprotected chain is typically hydrophilic. Fortunately, most of the common protecting groups are removed without difficulty, and deprotection levels greater than 97% are readily attained. The more serious consequence of deprotection is cleavage of the peptide chain, or racemization of the optically pure amino acid residues.

Basic polypeptides, such as polylysine or polyarginine, are readily deprotected without much difficulty [12, 13, 49]. Acidic polypeptides, such as polyglutamic acid or polyaspartic acid, require more care in deprotection reactions due to an abundance of potential side-reactions. PBLG, for example, can be debenzylated using strong acid, aqueous base, or catalytic hydrogenation. Strong acid (e.g. gaseous HBr or HBr in acetic acid) avoids any racemization, but is known to lead to significant chain cleavage arising from protonation of side-chain ester groups that react with the amide backbone [50]. Basic conditions avoid this reaction, but can lead to significant racemization unless the amount of base is carefully controlled [51, 52]. Hydrogenation would appear to be the most attractive method, however, it is only effective for chains less than 10 kDa in mass. Larger PBLG chains adopt a stable helical conformation that prevents access of the hydrogenation catalyst to the ester groups [51, 52]. Ester cleavage using trimethylsilyl iodide was found to give clean conversion to the readily hydrolyzed trimethylsilyl ester, without any racemization or chain cleavage [53]. The major drawbacks of this reagent are its expense as well as its reactivity with most other functional groups, such as the ether linkages in PEG. The deprotection of poly(β-benzyl-L-aspartate), PBLA, shows less side reactions under acidic conditions compared to PBLG. However, it has been reported that the polymer backbone undergoes partial rearrangement to β-peptide linkages under basic conditions, presumably through an imide intermediate [54, 55]. The degree of racemization in these samples was not discussed.

5
Conclusions

The synthesis of polypeptide hybrid block copolymers is an area that has been under study for three decades. Initially, this field suffered from limitations in the synthesis of the polypeptide components that required excessive sample purification and fractionation to obtain well-defined copolymers. In recent years, vast improvements in NCA polymerizations now allow the synthesis of hybrid block copolymers of controlled dimensions (molecular weight, sequence, composition, and molecular weight distribution). Such well-defined materials will greatly assist in the identification of new self-assembled structures possible using ordered polypeptide segments, as well as yield new materials with a wide range of tunable properties.

References

1. Branden C, Tooze J (1991) Introduction to Protein Structure. Garland, New York
2. Cha JN, Stucky GD, Morse DE, Deming TJ (2000) Nature 403:289
3. van Hest JCM, Tirrell DA (2001) Chem Commun 1897
4. Voet D, Voet JG (1995) Biochemistry, 2nd edn, chap 32. Wiley, New York
5. Fasman GD (1967) Poly α-Amino Acids. Dekker, New York
6. Fasman GD (1989) Prediction of Protein Structure and the Principles of Protein Conformation. Plenum Press, New York, p 48
7. Deming TJ (2000) J Polym Sci Polym Chem Ed 38:3011
8. Gallot B (1996) Prog Polym Sci 21:1035
9. Schlaad H, Antonietti M (2003) Eur Phys J E 10:17
10. Discher DE, Eisenberg A (2002) Science 297:967
11. Kwon GS, Naito M, Kataoka K, Yokoyama M, Sakurai Y, Okano T (1994) Colloid Surfaces B 2:429
12. Kricheldorf HR (1987) α-Aminoacid-N-Carboxyanhydrides and Related Materials. Springer, Berlin Heidelberg New York
13. Kricheldorf HR (1990) In: Penczek S (ed) Models of Biopolymers by Ring-Opening Polymerization. CRC Press, Boca Raton, FL
14. Woodward RB, Schramm CH (1947) J Am Chem Soc 69:1551
15. Webster O (1991) Science 251:887
16. Cardinaux F, Howard JC, Taylor GT, Scheraga HA (1977) Biopolymers 16:2005

17. Howard JC, Cardinaux F, Scheraga HA (1977) Biopolymers 16:2029
18. Kubota S, Fasman GD (1975) Biopolymers 14:605
19. Sekiguchi H (1981) Pure Appl Chem 53:1689
20. Sekiguchi H, Froyer G (1975) J Polym Sci Symp 52:157
21. Collman JP, Hegedus LS, Norton JR, Finke RG (1987) Principles and Applications of Organotransition Metal Chemistry 2nd edn. University Science Books, Mill Valley, CA
22. Deming TJ (1997) Nature 390:386
23. Deming TJ (1998) J Am Chem Soc 120:4240
24. Deming TJ (1999) Macromolecules 32:4500
25. Deming TJ, Curtin SA (2000) J Am Chem Soc 122:5710
26. Deming TJ (2002) Adv Drug Deliv Rev 54:1145
27. Deming TJ (2005) Soft Matter 1:28
28. Aliferis T, Iatrou H, Hadjichristidis N (2004) Biomacromolecules 5:1653
29. Vayaboury W, Giani O, Cottet H, Deratani A, Schué F (2004) Macromol Rapid Commun 25:1221
30. Odian G (1991) Principles of Polymerization 3rd edn. Wiley, New York
31. Dimitrov I, Schlaad H (2003) Chem Commun 2944
32. Knobler Y, Bittner S, Frankel M (1964) J Chem Soc 3941
33. Knobler Y, Bittner S, Virov D, Frankel M (1963) J Chem Soc C 1821
34. Fischer H (2001) Chem Rev 101:3581
35. Perly B, Douy A, Gallot B (1974) CR Acad Sci Paris 279C:1109
36. Yamashita Y, Iwaya Y, Ito K (1975) Makromol Chem 176:1207
37. Aoi K, Hatanaka T, Tsutsumiuchi K, Okada M, Imae T (1999) Macromol Rapid Commun 20:378
38. Maeda M, Inoue S (1981) Makromol Chem Rapid Commun 2:537
39. Imanishi Y (1984) J Macromol Sci Chem A21:1137
40. Uchida S, Oohori T, Suzuki M, Shirai H (1999) J Polym Sci Polym Chem 37:383
41. Schlaad H, Smarsly B, Losik M (2004) Macromolecules 37:2210
42. Lutz J-F, Schütt D, Kubowicz S (2005) Macromol Rapid Commun 26:23
43. Curtin SA, Deming TJ (1999) J Am Chem Soc 121:7427
44. Witte P, Menzel H (2004) Macromol Chem Phys 205:1735
45. Brzezinska KR, Deming TJ (2001) Macromolecules 34:4348
46. Brzezinska KR, Deming TJ (2004) Macromol Biosci 4:566
47. Brzezinska KR, Curtin SA, Deming TJ (2002) Macromolecules 35:2970
48. Kros A, Jesse W, Metselaar GA, Cornelissen JJLM (2005) Angew Chem Int Ed 44:4349
49. Ben-Ishai D, Berger A (1952) J Chem Soc 1564
50. Blout ER, Idelson M (1956) J Am Chem Soc 78:497
51. Hanby WE, Waley SG, Watson J (1948) Nature 161:132
52. Hanby WE, Waley SG, Watson J (1950) J Chem Soc 3239
53. Subramanian G, Hjelm RP, Deming TJ, Smith GS, Li Y, Safinya CR (2000) J Am Chem Soc 122:26
54. Saudek V, Pivcová H, Drobník J (1981) Biopolymers 20:1615
55. Yokoyama M, Kwon GS, Okano T, Sakurai Y, Seto T, Kataoka K (1992) Bioconjugate Chem 3:295
56. Tsutsumiuchi K, Aoi K, Okada M (1997) Macromolecules 30:4013
57. Kim KT, Vandermeulen GWM, Winnik MA, Manners I (2005) Macromolecules 38:4958
58. Kong X, Jenekhe SA (2004) Macromolecules 37:8180
59. Kricheldorf HR, Hauser K (2001) Biomacromolecules 2:1110

Adv Polym Sci (2006) 202: 19–52
DOI 10.1007/12_081
© Springer-Verlag Berlin Heidelberg 2006
Published online: 23 February 2006

Synthesis of Bio-Inspired Hybrid Polymers Using Peptide Synthesis and Protein Engineering

Dennis W. P. M. Löwik · Lee Ayres · Jurgen M. Smeenk ·
Jan C. M. Van Hest (✉)

Organic Chemistry Department, Institute for Molecules and Materials,
Radboud University Nijmegen, Toernooiveld 1, 6525 ED Nijmegen, The Netherlands
j.vanhest@science.ru.nl

Abstract The construction of well-defined hybrid materials consisting of synthetic polymers and (poly)peptides or proteins has attracted much attention in recent years. Different techniques have become available that allow an efficient synthesis of polymer architectures with (poly)peptides in the side chain as well as the main chain. Oligopeptides modified with monomer or initiator moieties can, for example, be introduced in polymers via controlled (radical) polymerization methods. Ligation methods enable the build up of large peptide structures via a complete organic chemistry approach. The materials scientist's toolbox is enriched with molecular biology techniques, and protein engineering has become a standard synthetic method for the construction of well-defined polypeptides with specific functional handles for the attachment of synthetic polymers. The availability of all these techniques holds much promise for the application of materials in the biomedical field, and especially tissue engineering, targeted drug delivery and diagnostics should in the near future benefit from these recent synthetic achievements.

Keywords Hybrid polymers · Peptides · Proteins · Controlled polymerization · Conjugation

1
Introduction

Polymer-peptide hybrid architectures have recently gained much interest in materials science. The ability to combine the structural and functional control of proteins with the versatility of synthetic polymers has given access to a novel class of molecules that can find their application in areas as diverse as nanotechnology, drug delivery and tissue engineering [1–4]. The increased activities in bio-hybrid materials science can be explained for a large part by the development of new synthetic methodologies that allow for the build up of well-defined polymer–protein hybrid materials [5, 6]. Techniques from organic and peptide chemistry, polymer chemistry and molecular biology have enriched the materials scientist's toolbox.

The combination of solid phase peptide synthesis with polymer chemistry has proven to be a versatile method for the preparation of polymer-peptide hybrids. Introduction of native ligation methods even allows the synthesis of polymer modified polypeptides and proteins via an entire organic chemistry approach. In the field of polymer chemistry—besides the advances in NCA polymerization, which will be discussed by others and is therefore not part of the scope of this review—controlled radical polymerization has been shown to be a robust technique, capable of creating well-defined biofunctional polymer architectures. Through protein engineering, methods have been established that enable the construction of tailor-made proteins, which can be functionalized with synthetic polymer chains in a highly defined manner.

There are a number of applications in which it is advantageous to use hybrid materials. Proteins and peptides are easily denatured by proteases, heat and adverse solvent conditions, and when introduced into the body they can sometimes generate negative (immune) responses. The use of a peptide polymer-hybrid material can help to reduce these problems. A polymer coupled to a protein can shield the protein from any unwanted interactions. Including only the active part of a protein in a polymer and removing the rest of the protein could also help to reduce any undesired biological response. Thus, by preparing peptide polymer hybrid materials it is possible to use functional proteins or peptides throughout a wider range of biomedical applications and conditions [7, 8].

Peptide-polymer hybrid materials can also be used to more efficiently prepare (biologically) active coatings. When an active peptide or protein is included in a coating using only physisorption, leaching can easily occur. This leads to a lowered surface activity and it creates the possibility of a negative biological response to the free peptide. Anchoring the desired peptide or protein to the surface by creating a peptide-polymer hybrid material stops this leaching process, thereby removing unwanted biological side effects and increasing the active lifetime of the coating or surface [9].

Another advantage of using peptide-polymer hybrid materials is that multiple functionalities can be included in the same polymer chain, or can be brought together in the same aggregate. This can be achieved by creating (block) copolymers of different peptide sequences or by connecting different proteins to polymer chains to create large amphiphiles which can self assemble into functional aggregates [10].

In this review the different synthetic strategies for constructing polymer-peptide hybrids will be discussed, as well as some of the characteristic features of the materials. The complexity of the hybrid structures prepared will increase as this review progresses, starting with the controlled polymerization of peptide-containing monomers and later covering the creation of block copolymer structures via a protein engineering approach.

2
Polymers Containing Side Chain Peptide Moieties

2.1
Grafting Methods

One of the earliest approaches to creating polymer-peptide hybrid materials was based on the introduction of peptide moieties in the polymer side chain [11]. These architectures have been of particular interest for the development of drug delivery systems with a high loading capacity as well as for systems in which case the bioactivity is related to having multiple copies of a peptide in close proximity of each other.

There are in general two ways to synthesize side chain polymers, polymerization of peptide-functional monomers or introduction of the peptide moiety afterwards, by grafting. The latter technique is based on the synthesis of polymers containing some form of functionality in the side chain, normally an activated ester moiety, which can further react with a peptide. The most commonly used method for the polymerization of monomers containing active esters is free radical polymerization. In particular many activated acrylate esters have been polymerized in this manner [12] (Table 1) for use in a wide variety of applications, from the preparation of polymer drug conjugates [13, 14] to supports for solid phase peptide synthesis [15, 16].

Although free radical polymerizations are facile, they do not yield well-defined polymer architectures. Good control over the polymerization, however, is essential to allow the synthesis of more complex architectures which might improve the activity of the polymer peptide hybrid.

To create better-defined polymers suitable for peptide grafting, Godwin et al. [17] chose to use the controlled radical polymerization method Atom Transfer Radical Polymerization (ATRP) to polymerize N-hydroxy succinimide methyl acrylate (NHSM). They then coupled glycyl-glycine-β-naphthylamide to the side chain of this polymer, creating a well-defined peptide-polymer hybrid material in a straightforward manner. Schilli and co-workers used in more recent investigations reversible addition–fragmentation chain transfer (RAFT), another type of controlled living polymerization technique, to polymerize monomers containing an activated ester moiety [18]. They not only polymerized NHSM but also a wider range of monomers, including 2-vinyl-4,4-dimethyl-5-oxazolone (VO), diacetone acrylamide (DAA), N-isopropyl acrylamide (NIPAAm) and acrylic acid (AA) (Fig. 1). The conjugation of these polymers to model peptides was investigated.

Theato et al. [19] studied the polymerization of activated esters such as 2,4,5-trichlorophenyl acrylate and endo-N-hydroxy-5-norbornene-2,3-dicarboxyimide acrylate (NORB, Fig. 1) using ATRP. They showed that the polymer side chains were reactive by performing a coupling reaction with ammonia.

Another approach, which does not make use of either free or controlled radical polymerization, was demonstrated by Parrish et al. [20]. An aliphatic polyester with pendent acetylene groups was prepared via controlled ring-opening polymerization. Polyethylene glycol and the peptide sequence Gly – Arg – Gly – Asp – Ser (GRGDS) were functionalized with an azide moiety, and subsequently "clicked" to the pendent acetylenes in the

Table 1 Activated acrylates (AOR represents $CH_2 = CHCO - OR$) reported for the synthesis of carboxyl-activated polymer intermediates

Monomer	R
AOBt	1-benzotriazole [112]
AOSu	N-succinimide [15, 112]
AOPcp	pentachloro phenyl [14]
AOTcp	2,4,5-Trichlorophenyl [15, 112]
AONp	4-nitrophenyl
AOQu	8-quinoline [15]
AOPy	3-pyridyl [15]
AOCp	2-carboxyphenyl [15]
AOMcp	2-methoxycarbonylphenyl [15]
AOCl	chloride [113]

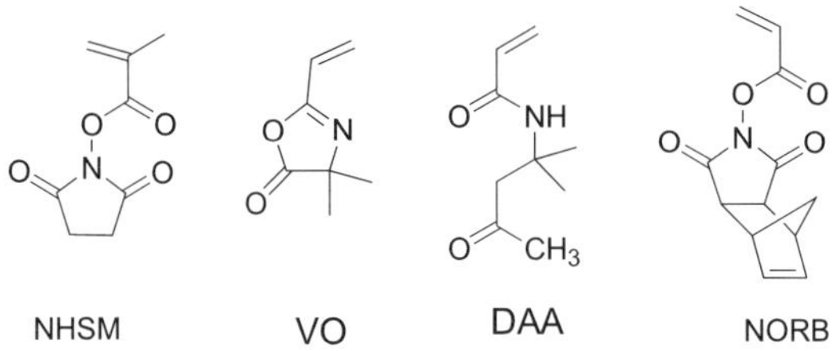

NHSM   VO   DAA   NORB

Fig. 1 Different types of functional monomers used to create polymers with grafting functionality via controlled radical polymerization. The abbreviations correlate with the ones used in the text [20, 21]

side chain of the polymer using Cu(I) mediated 1,3 Huisgen dipolar cycloadditions. In vitro cytotoxicity evaluation showed these polymers to be biocompatible.

2.2
Polymerization of Peptide-Containing Monomers

2.2.1
Free Radical Polymerization

The major drawback to using the grafting approach is that it is very difficult to achieve and/or determine 100% functionalization of the polymer side chains. To overcome this problem, another approach has been developed, in which the monomer already contains the peptide fragment of interest. Therefore, after polymerization, every monomer unit is inherently functionalized [21]. The disadvantage of this method is that synthesizing peptide-based monomers is not trivial, and compatibility issues between polymerization method and peptide moiety have to be taken into account.

Again free radical polymerization has been used by a large number of researchers to polymerize amino acid-containing monomers [22]. Endo and co-workers for example extensively investigated this preparation method by synthesizing several monomers based on leucyl-alanyl repeats and leucyl-alanylglycine moieties. Peptides were either directly functionalized with methacrylic acid to yield the methacryl amide derivative, or peptides were coupled to hydroxy ethyl methacrylate (HEMA) to yield the methacrylate monomers with an ethyl spacer between the peptide and the polymerization handle. The effect of peptide length on free radical polymerization was examined by polymerizing methacrylamide functional oligopeptides containing 1 to 4 repeats of the leucine–alanine sequence [23, 24]. It was observed that all peptides could be conveniently polymerized except for the octa-peptide, which only resulted in a degree of polymerization of 3.5. This lower polymerizability was tentatively attributed to the aggregation of the octa-peptide monomer, due to hydrogen bonding.

Kopeček and co-workers [25, 26] used free radical polymerization chemistry to construct polymer-drug conjugates as anti-tumor agents. They prepared (biocompatible) copolymers of N-(2-hydroxypropyl)-methacrylamide and a methacrylamide-functionalized peptide, such as Gly – Phe – Leu – Gly (GFLG), containing a p-nitroaniline group on the carboxylic acid terminus. In the next step it was possible to graft the anti-cancer drug doxorubicin onto the active ester of the peptide (Fig. 2). The connection of a drug to a polymer carrier resulted in all the normal characteristic advantages of polymer drug delivery systems. First, the toxicity effects of doxorubicin were drastically decreased, the circulation time of the conjugate within the body was increased compared to the unbound drug by a factor of five, and furthermore, the polymeric drug showed a higher level of accumulation in tumor tissue than in healthy tissue, due to the well-documented enhanced permeation and retention (EPR) effect. The additional benefit of introducing the GFLG peptide spacer between polymer and doxorubicin was that this sequence was specifically cleaved by enzymes present in the lysosome, which resulted in controlled drug release. A more specific polymer variant was made by the combined introduction of doxorubicin and N-acylated galactosamine, of which the latter

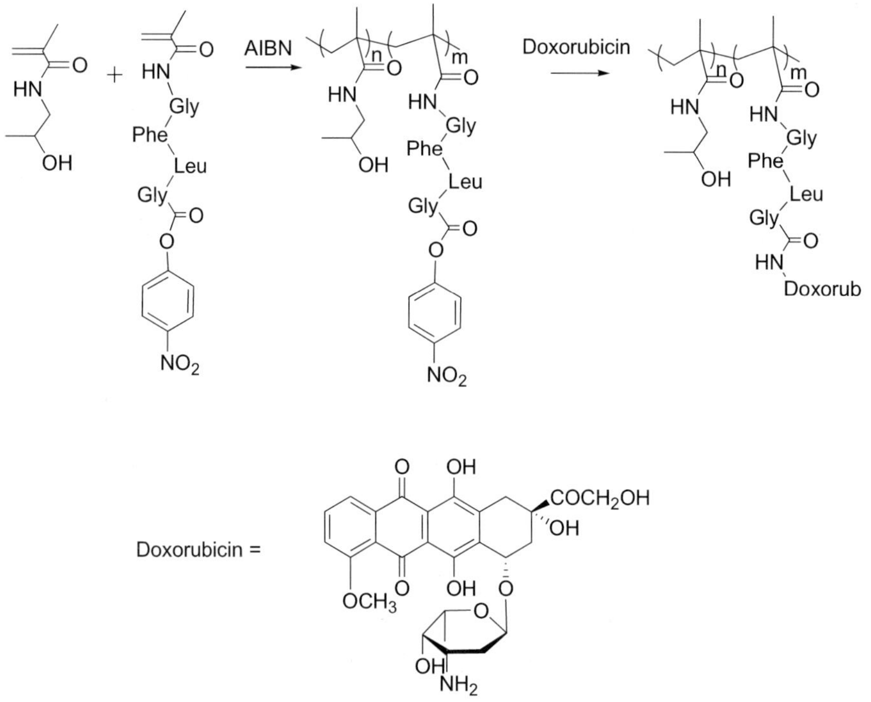

Fig. 2 Preparation of a polymer-drug conjugate from the copolymerization of N-(2-hydroxypropyl)-methacrylamide and a methacrylamide-functionalized peptide-based monomer [27]

was used for targeting protein receptors in liver cells. These conjugates are the first examples of peptide-polymer hybrid materials used as drug delivery agents, and are currently under investigation in clinical trials.

Side chain peptide polymers are especially of interest in the development of synthetic vaccines. The preparation of defined, chemically synthesized peptide-antigens could exclude the use of infectious material and thus result in safer vaccines. However, short peptides generally do not elicit good immune responses, therefore several synthetic approaches have been used to generate higher molecular weight polymers from them [27]. An interesting approach has been described using acryloyl-modified peptides, which were polymerized at room temperature by free radical polymerization in a buffered aqueous solution. In order to prevent steric interactions of the peptides copolymerizations were performed with a typical 50 molar excess of acrylamide. This method allowed the successful production of copolymers of different epitopes, varying in length from 4 to 23 amino acid residues [28]. In another paper by Jackson et al. [29] it could be shown from antigen-antibody interaction inhibition studies that these polymer epitopes were more antigenic than their monomeric variants.

2.2.2
Metathesis Polymerization

One of the first examples of the polymerization of a peptide-based monomer via a more controlled technique was published by Maynard and Grubbs [10]. They prepared a series of norbornenes functionalized with cell adhesion peptides such as GRGDS and polymerized them using ring opening metathesis polymerization (ROMP), yielding polymers with an M_n around 10 kg/mol and a PDI in the order of 1.3. It was investigated how the prepared polymers inhibited cell adhesion to fibronectin, a natural protein containing GRDGS, which can be taken as a measure of the binding affinity of the polymers to cells. It was observed that by including the peptide sequence GRGDS in the side chain of a polymer inhibition was increased when compared to the free

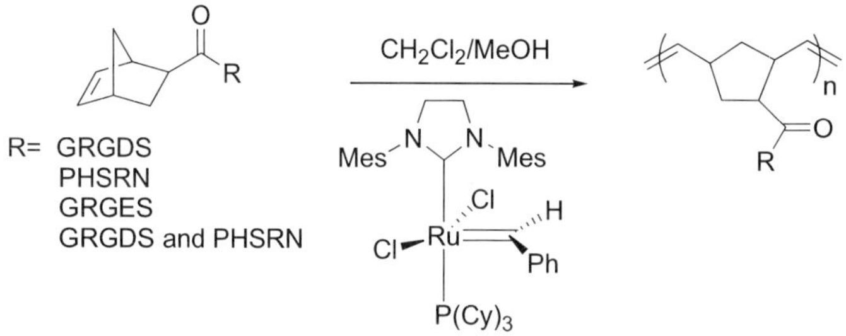

Fig. 3 Synthesis of homo- and copolymers with pendent bioactive oligopeptides by Ring Opening Metathesis Polymerization (ROMP). E = Glu (glutamic acid) [10]

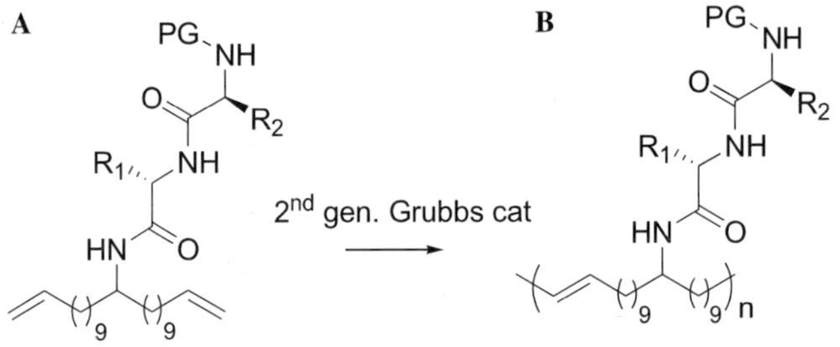

Fig. 4 Polymerization of 2 dipeptide functional dienes by ADMET. A R_1 = CH(CH$_3$)$_2$, R_2 = CH$_3$, PG is protecting group (CBz); B R_1 = R_2 = CH$_2$CH(CH$_3$)$_2$, PG = Boc

peptide. The authors proposed that the conjugation of the peptide to the polymer increased the local concentration of the peptide, thereby enhancing the number of peptide-cell interactions. The increased concentration effect highlights one of the advantages of including peptides in the side chain instead of the main chain of the polymer.

This line of research was extended by preparing random copolymers, which contained besides GRGDS a second peptide, Pro – His – Ser – Arg – Asn (PHSRN). PHSRN is a synergistic peptide which increases the binding potential of GRGDS in fibronectin. By conjugating both peptides to the same polymer backbone they are brought into close proximity, as is the case with fibronectin, and the synergistic effect can be optimally utilized. This was indeed demonstrated with the abovementioned binding inhibition assay. Besides ROMP also acyclic diene metathesis polymerization (ADMET) has been applied for the construction of peptide-containing polymers [30]. Amino acid- and dipeptide-functionalized branched dienes (Fig. 4) were synthesized and polymerized to yield chiral polyolefins. Polymer A proved to be crystalline with a melting temperature of 150 °C, whereas polymer B was amorphous. This different behavior was explained both from the use of a different protecting group as well as from the presence of more hydrophobic side chains in the case of polymer B.

2.2.3
Isocyanide Polymerization

Isocyanide polymerization has also been used to polymerize peptide-based monomers. Cornelissen et al. [31, 32] prepared oligopeptides based on alanine and functionalized the N-terminus with an isocyanide moiety. These monomers were subsequently polymerized using a Ni catalyst into β-helical poly isocyanopeptides with the dipeptides in the side chain. It was found that these polymers formed rigid rods, which were revealed by AFM to have extremely long persistence lengths. This rigidity was caused by the formation of β-sheets between the alanines in the side chain. The same group

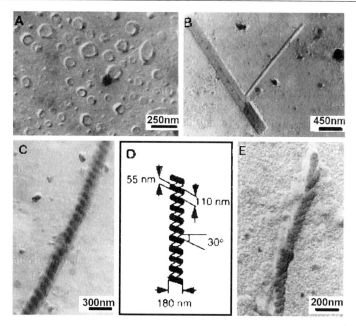

Fig. 5 TEM micrographs of the morphologies formed by polystyrene-block-poly(iso-cyanodipeptide)s in aqueous solution: **A** vesicles, **B** bilayer filaments, **C** left-handed super-helix, **D** schematic representation, and **E** right-handed helical aggregate. Reprinted with permission from [33]. Copyright 1998 AAAS

also investigated amphiphilic block copolymers containing a poly(styrene) tail and a charged poly(isocyanide) headgroup, derived from isocyano-L-alanine-L-alanine and isocyano-L-alanine-L-histidine. This type of rod-coil block copolymers formed micelles, vesicles and bilayer aggregates in aqueous solution as is depicted in Fig. 5. Furthermore, the chirality of the poly(isocyanodipeptide) block was shown to be dependent on the dipeptide used. Moreover, helical superstructures were of opposite chirality to that of the constituent block copolymer [33–35].

2.2.4
Acetylene Polymerization

Tang and co-workers used leucine-functionalized phenyl acetylene derivatives for the construction of amphiphilic helical polymers, which were envisioned to be both semi-conducting and biocompatible, leading to diverse applications such as biosensors [36]. The polymerization was performed with a rhodium catalyst and resulted in high molecular weight polymers, particularly for polyacetylene **1a** ($1.5 \cdot 10^6$ g/mol). Interestingly, only the polymers in which the stereo-center was closely located to the helical backbone (**1a** and **1b**) showed a CD signal and were optically active (Fig. 6).

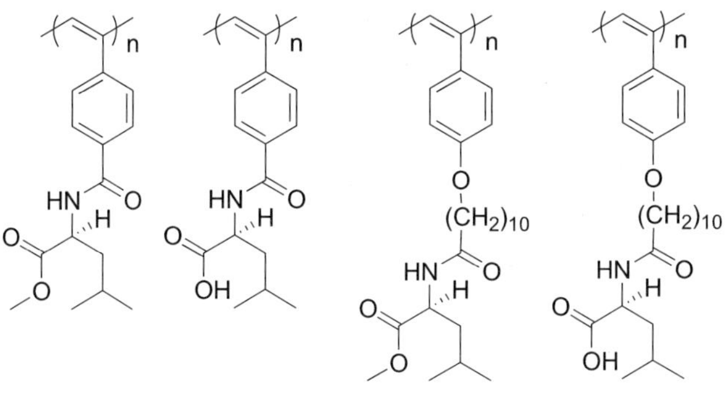

Fig. 6 Phenyl acetylene polymers containing amino acid side chains [36]

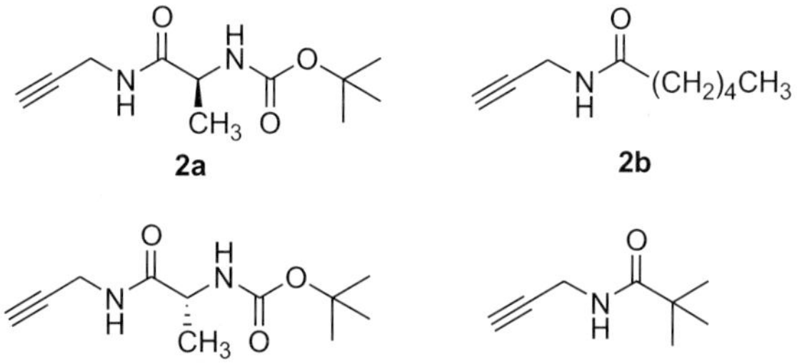

Fig. 7 Acetylene-functionalized peptides. Inclusion of only 12% of the L and D chiral monomers (2a and 2c) into the polymerization of achiral monomers 2b and 2d result in the same helix content as for the homopolymers of 2a and 2c [37]

In a similar line of research Gao and co-workers [37] prepared helical polymers by copolymerizing acetylene functionalized D and L alanine with the achiral hexanoic acid N-propargylamide and pivalic acid N-propargylamide (Fig. 7), again using a rhodium-based catalytic system. They demonstrated that including as little as 12% of the chiral peptide-based monomer still resulted in a polymer with the same helical content as the peptide-based homopolymer. The same group also prepared threonine- and serine-based polyacetylenes, of which the helical content could be increased by changing solvent from THF to MeOH [38]. It was proposed that the presence of the free OH groups in the peptide side chains were responsible for this behavior.

2.2.5
Controlled Radical Polymerization

The first example of the use of a controlled radical polymerization technique for the construction of polymers with side chain peptide moieties was demonstrated by Ayres et al. Atom transfer radical polymerization (ATRP) was used for the preparation of side chain polymers containing pentapeptide repeats commonly found in tropoelastin (Val – Pro – Gly – Val – Gly, (VPGVG)) [39]. Elastin is a structural protein responsible for the elastic behavior of mammalian tissue. The uncrosslinked precursor, tropoelastin, exhibits a specific lower critical solution (LCST) behavior, which results in aggregation upon temperature increase and resolubilization upon cooling. It has been reported that this feature is an intrinsic property of one pentapeptide repeat [40]. Block copolymers with pentapeptide elastin side chains were therefore envisaged to elicit thermo-responsive behavior. The pentapeptide monomers were synthesized either by solution or solid phase peptide chemistry, followed by modification of the N-terminus with a methacrylate handle. Subsequent ATRP from a bifunctional poly(ethylene glycol) macroinitiator resulted in a series of triblock copolymers with a narrow molecular weight distribution (PDI = 1.2), and a controllable degree of polymerization. These block copolymers showed a DP-, temperature- and pH-dependent aggregation behavior which was indeed similar to elastin [41].

The same methodology was applied to construct side chain polymers containing the beta-sheet forming tetrapeptide Ala – Gly – Ala – Gly (AGAG) [42]. This side chain polymer was subsequently used as a macroinitiator for ATRP to prepare flanking blocks of amorphous poly methyl methacrylate. Infra-red spectroscopy clearly showed that the resulting well-defined triblock copolymer possessed a beta-sheet secondary structure.

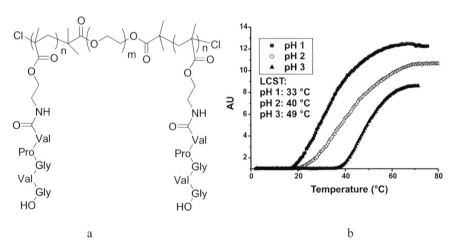

a   b

Fig. 8 pH-dependent LCST behavior of a triblock copolymer containing elastin side chains. **a** Chemical structure; **b** turbidity measurements performed at different pH (1, 2 and 3) for a triblock copolymer consisting of a PEG block of M_n = 1000 g/mol and elastin blocks with n = 11. Reprinted with permission from [41]. Copyright 2005 American Chemical Society

3
Main Chain Peptide-Polymer Hybrids

3.1
Coupling of Pre-Made Polymers to Peptide Moieties

3.1.1
Solution Phase Peptide Synthesis

Solution phase synthesis of peptides is useful for the preparation of short sequences of up to 10 residues in length. Difficulties that limit this technique
are non-quantitative coupling reactions and the often problematic solubility
of oligopeptides. Only a few examples have been reported in which small peptidic elements prepared by solution phase peptide synthesis have been used
in the construction of "hybrid" multiblock copolymers. An interesting example in which the peptide segment was used to contribute to the mechanical
properties of the final material is the work of Sogah et al. [43–45]. They prepared *Bombyx mori* silk worm silk and *Nephila clavipes* spider dragline silk
mimetic block copolymers. Silk worm silk is of interest because of its use in
textile fibers, whereas spider dragline silk is one of the types of spider silk
known to have a remarkable combination of strength and toughness [46].
Both silks contain repetitive amino acid sequences which form crystalline
and amorphous domains in the silk fiber. The crystalline domains give the
material strength, whereas the amorphous protein matrix allows the crystalline domains to orient under strain to increase the strength and the energy
needed to snap the material by the introduction of flexibility. The crystalline
β-sheet sequences are poly(alanylglycine) in silkworm silk and poly(alanine)
in dragline silk. Sogah et al. used either the tetrapeptide Gly – Ala – Gly – Ala
or poly(Ala) and replaced the amorphous silk sequences with poly(ethylene
glycol). Multiblock copolymers were prepared via polycondensation using
two approaches. The first method made use of an aromatic hairpin residue
to force formation of parallel β-sheets as is schematically depicted in Fig. 9a.
A second approach is depicted in Fig. 9b and is a fully linear system in which
the β-sheet segments were free to form intra- and intermolecular parallel or
antiparallel β-sheets.

For both designs a microphase separated morphology was found with
20–50 nm peptide domains dispersed in a continuous poly(ethylene glycol)
phase. Furthermore, a 100–150 nm superstructure was observed in cast films,
which was explained to result from the polydispersity and multiblock character of the polymers. The mechanical properties of fibers and films made
from these block copolymers could be modulated by manipulating the length
and nature of the constituent blocks. Similar work was reported by Shao
et al. [47].

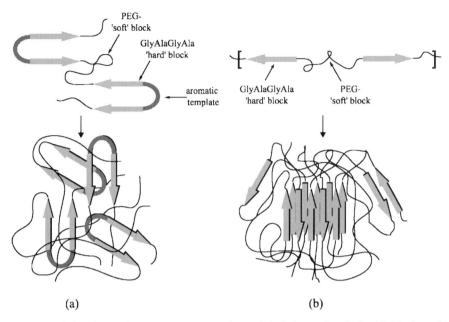

Fig. 9 Multiblock copolymers consisting of a poly(ethylene glycol) "soft" block and a tetrapeptide Ala-Gly-Ala-Gly, crystalline "hard" block in two variants: **a** Templated system in which an aromatic hairpin turn is used to force parallel β-sheet formation. **b** Non-templated system in which peptide segments are free to form parallel and/or antiparallel β-sheets. Reprinted with permission from [43]. Copyright 2001 American Chemical Society

3.1.2
Solid Phase Peptide Synthesis

For the synthesis of small to medium-sized peptides solid phase peptide synthesis is the method of choice. This technique involves the stepwise addition of *N*-protected amino acids to a peptide chain anchored with its C-terminus to a polymeric support [48]. With this method a peptide is constructed via sequential coupling and deprotection steps and thus sequence-specific peptides can be prepared. The immobilization of the peptide allows the use of excess of coupling and deprotection reagents, resulting in high yields per step. Kopeček and co-workers [49] used solid phase peptide chemistry to produce repeats of the oligopeptide Val – Ser – Ser – Leu – Glu – Ser – Lys, (VSSLESK)$_n$, where n = 2,3,4 and 5. This peptide sequence is well known to form coiled coil motifs. They then coupled these peptides to a PEG chain via standard amide chemistry, and demonstrated that coiled coil association indeed occurred, and that this coiled coil interaction was stabilized through conjugation with PEG. Klok et al. [50] used a similar approach to produce an alternative coiled coil peptide sequence, Gly – (Glu – Ala – Lys – Leu – Ala – Glu – Ile)$_3$ – Tyr

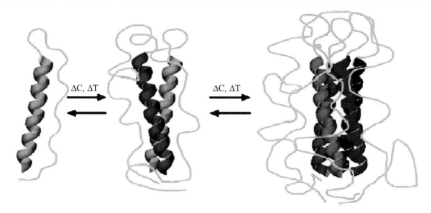

Fig. 10 Proposed model for the self assembly of PEG-coiled-coil hybrid polymers, which is dependent on both concentration and temperature. Reprinted with permission from [50]. Copyright 2003 American Chemical Society

$(G(EAKLAEI)_3Y)$ functionalized with a PEG chain. They demonstrated that the number of coiled coils within an aggregate decreased with increasing PEG chain length, due to steric interference by PEG. Interestingly, the main driving force behind the formation of these supramolecular aggregates is not selective solvation, as known from classical amphiphilic block copolymer aggregation, but rather the entropy-driven hydrophobic interaction between the coiled coil peptide blocks.

Rossler et al. used the same method to prepare β-strand peptide sequences based on lysine and leucine repeats. These peptides were coupled to poly ethylene glycol [51] leading to architectures similar to those prepared by Sogah, however, instead of preparing multiblock copolymers, triblock copolymers were produced. They investigated the secondary structure by infrared spectroscopy and X-ray scattering and confirmed that the peptide strands aggregated into β-sheets, leading to a well-defined nanostructured material.

Reynhout et al. [52] recently published a completely solid phase method for the synthesis of peptide-based triblock copolymers. Amine-functionalized polystyrene was first coupled to an aldehyde-modified resin to give a polystyrene functionalized secondary amine on the resin, which could then be coupled to the next amino acid (Fig. 11). Then the sequence Gly – Ala – Asn – Pro – Asn – Ala – Ala – Gly (GANPNAAG)—a known β-hairpin folding peptide found in the CS protein of the Malaria parasite *plasmodium falciparum*—was synthesized, using standard Fmoc peptide chemistry. After removing the final Fmoc group from the peptide, a carboxylic acid-functionalized polystyrene was coupled to the terminal amine.

The aggregation behavior of this amphiphilic block copolymer was investigated using electron microscopy, and spherical aggregates of around 250 nm in diameter were observed.

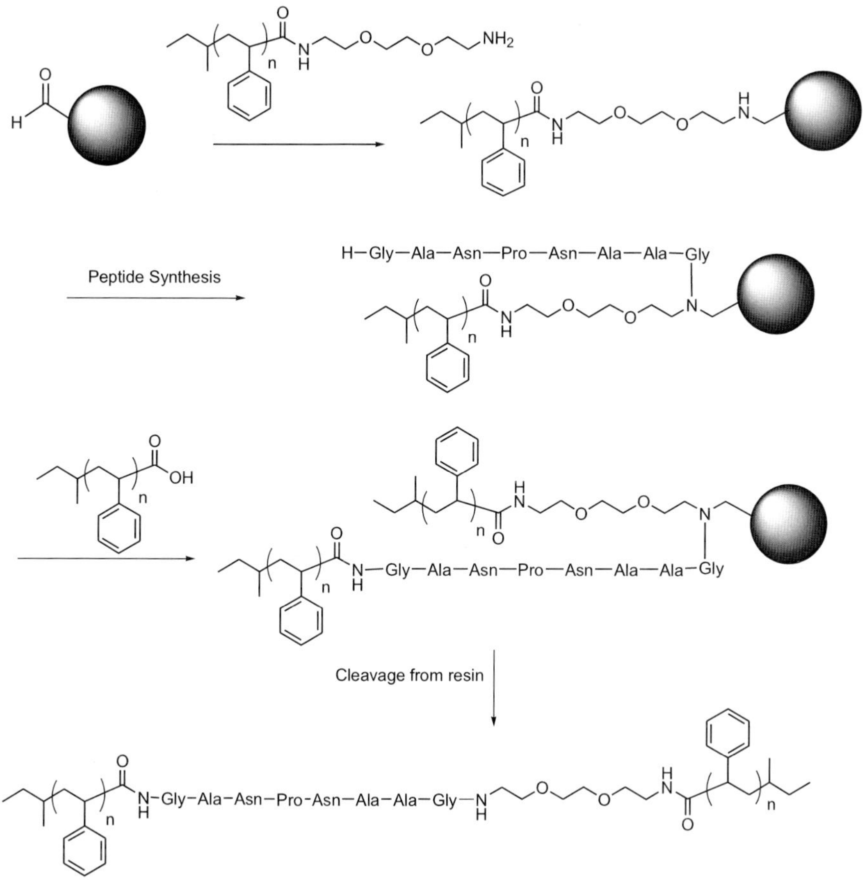

Fig. 11 Totally solid phase approach to the synthesis of a peptide-polymer hybrid material [52]

Another interesting example was reported by Burkoth et al. [53–55] who investigated fibril formation of β-amyloid peptide, the primary component of amyloid plaques in Alzheimer's disease. They derivatized the C-terminus of the central domain of β-amyloid peptide (Aβ_{10-35}) with poly(ethylene glycol)-3000 using standard Fmoc solid phase peptide synthesis. Although some models for β-amyloid structure (with the C-terminal hydrophobic domain in the fibril interior) suggested the attachment of PEG would disturb fibril formation, the conjugate did self-associate into fibrils. The fibril formation was, unlike the native peptide, completely reversible and the solubility of the formed fibrils was greatly improved. The lateral association of the fibrils into bundles was prevented, as can be seen in the electron micrographs of Fig. 12. Recently, a similar approach was described by Eckhardt et al. [56], who created polyethylene oxide–peptide block copolymers, in which the pep-

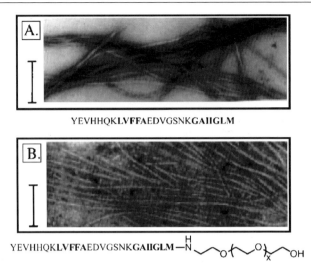

Fig. 12 Comparison of fibril morphology by electron microscopy. **A** β-amyloid peptide (Aβ(10–35)) and **B** Aβ(10–35)-PEG-3000 conjugate. Scale 200 nm. Reprinted with permission from [53]. Copyright 1998 American Chemical Society

tide element was designed to adopt a β-hairpin fold via the introduction of a β-turn mimetic template.

3.1.3
Ligation Methods

The advantage of using solid phase synthesis to prepare the desired peptide is that it is possible to have complete control over the peptide sequence. One of the main disadvantages, however, is that there is a limit to the length of the peptide which can be made without introducing too many deletions [57, 58]. To overcome this limitation new ligation methods have been developed. Native chemical ligation is based on the reaction of a peptide-α-thioester with another peptide segment containing an amino-terminal cysteine residue (Fig. 13a). The thioester-linked intermediate that is formed spontaneously rearranges, resulting in the formation of an amide bond and the regeneration of the free sulfhydryl group. An interesting practical extension of this concept is the fully solid-phase-based chemical ligation which allowed the build-up of longer peptides (e.g. up to 118 amino acids) in either the N → C or the C → N direction [59]. Furthermore, the development of the 1-phenyl-2-sulfanylethyl auxiliary group has allowed the ligation of peptides without the requirement of a cysteine group [60] (Fig. 13b). An extensive review on the chemical synthesis of proteins by ligation has been reported by Borgia et al. [61].

Elegant work in which the chemical ligation methodology was used for the preparation of a polymer-modified protein was reported by Kochendoerfer

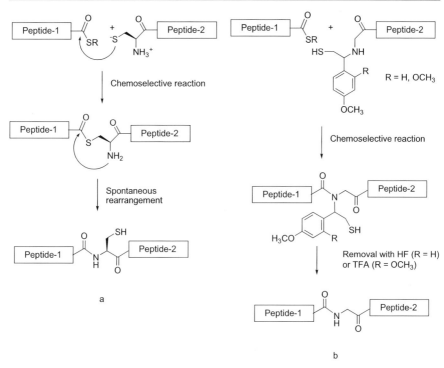

Fig. 13 a The principle of native chemical ligation. Peptide 1, containing a C-terminal thioester, undergoes a nucleophilic attack by the cysteine residue at the *N*-terminus of peptide 2. The intermediate rearranges spontaneously to form the native peptide bond. **b** Native chemical ligation without the requirement of a cysteine residue using the 1-phenyl-2-sulfanylethyl auxiliary group

et al. [62]. They prepared a monodisperse, polymer-modified, synthetic erythropoiesis protein (SEP). SEP was designed to be analogous to human erythropoietin (Epo), a glycoprotein hormone that regulates the proliferation, differentiation and maturation of erythroid cells. SEP was assembled by native chemical ligation of four individual peptide segments (from 27 to 55 amino acids) resulting in a polypeptide chain of 166 amino acids. Before this ligation, two of the peptide segments were functionalized via oxime-forming ligation [63] with a monodisperse, negatively charged and branched poly(ethylene glycol) derivative. These polymer moieties were attached to sites corresponding to two glycosylation sites of Epo and their negative charge enabled prolonged duration of action in vivo. This synthetic methodology resulted in a precisely defined 51 kDa protein-polymer conjugate that was produced in quantities of 100 mg, which is a considerable synthetic improvement in comparison to common protein-polymer conjugates that are often heterogeneous with respect to the polymers attached and the attachment sites.

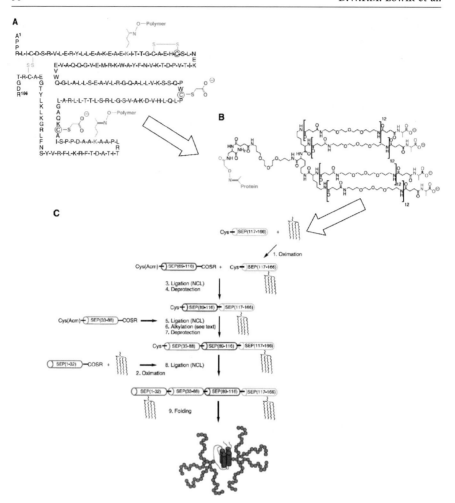

Fig. 14 Molecular structure of synthetic erythropoietin protein (SEP). **A** Primary amino acid sequence with three ligation sites circled in red. **B** Structure of the branched, negatively charged poly(ethylene glycol)-based polymer. **C** Scheme for the synthesis of SEP by chemical ligation. Branched polymers were first attached to the individual peptide segments by oxime-forming ligation followed by native chemical ligation. Reprinted with permission from [62]. Copyright 2003 AAAS

3.2
Peptide Synthesis Combined with Polymerization

Alternative total solid phase-based strategies for the preparation of polymer-peptide block copolymers were based on the polymerization of the synthetic polymer block from the supported peptide segment using either nitroxide-mediated radical polymerization (NMP) or ATRP (Fig. 15) [64, 65].

Becker et al. [64] functionalized a peptide, based on the protein transduction domain of the HIV protein TAT-1, with an NMP initiator while on the resin. They then used this to polymerize *t*-butyl acrylate, followed by methyl acrylate, to create a peptide-functionalized block copolymer. Traditional characterization of this triblock copolymer by gel permeation chromatography and MALDI-TOF mass spectroscopy was, however, complicated partly due to solubility problems. Therefore, characterization of this block copolymer was mainly limited to ^{1}H and ^{19}F NMR and no conclusive evidence on molecular weight distribution and homopolymer contaminants was obtained. Difficulties in control over polymer properties are to be expected, since polymerization off a microgel particle leads to a high concentration of reactive chains and a diffusion-limited access of the deactivator species. The traditional level of control of nitroxide-mediated radical polymerization, or any other type of controlled radical polymerization, will therefore not be straightforward to achieve.

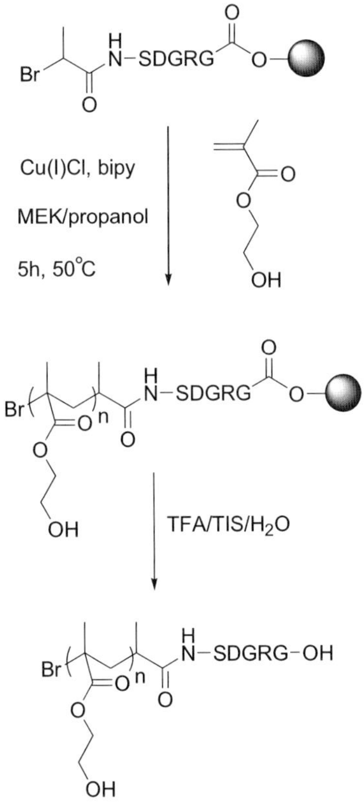

Fig. 15 Solid phase strategy towards peptide-containing block copolymers by polymerization from the supported peptide via atom transfer radical polymerization [66]

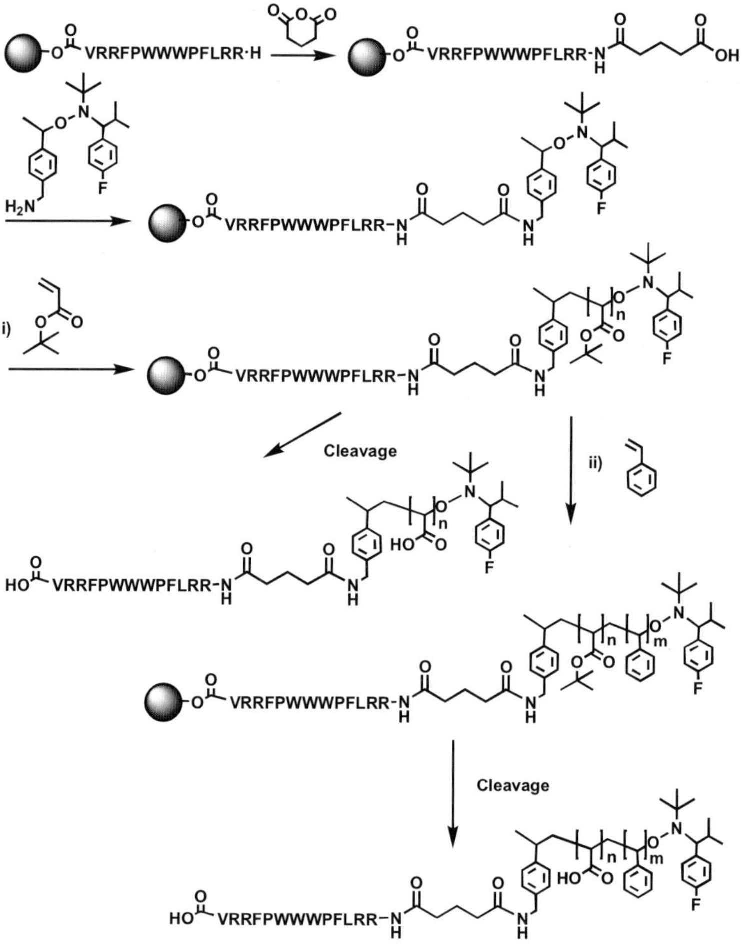

Fig. 16 The synthesis of a peptide polymer hybrid material from a tritrpticin-loaded solid support via nitroxide-mediated polymerization [66]

In a second article the same approach was used to synthesize a peptidic polymer containing the peptide Tritrpticin, a 13 residue antimicrobial peptide (Fig. 16) [66]. This time they initiated the polymerization of *t*-butyl acrylate, followed by styrene to produce a triblock copolymer, which clearly formed micelles in solution. Interestingly, the antimicrobial activity of the peptide was enhanced relative to the free peptide and the detrimental side effects normally associated with antimicrobial peptides, such as a high hemolytic activity, were reduced, highlighting the benefits of using peptide polymer hybrids in place of peptides alone.

The addition of an α-bromo ester or amide to the end of a molecule is a strategy which is commonly used to create a wide range of functional initiators for ATRP [67]. Rettig et al. [68] used this approach to develop a versatile method for the synthesis of a peptide-based main chain polymer. They synthesized a peptide using solid phase synthesis techniques and then functionalized it with an α-bromo amide. The peptide was cleaved from the resin and used to initiate the polymerization of n-butyl acrylate (nBA). They demonstrated that the polymers that were made had narrow molecular weight distributions, with PDI of 1.19, and that the architecture that was made was well defined.

A similar approach was followed by Couet et al. [69] who functionalized peptide Ghadiri rings on 3 sites with ATRP initiators, which were subsequently used for the polymerization of N-isopropyl acrylamide. These hybrid polymer peptide structures were capable of forming peptide nanotubes, which were homogeneously covered by the attached polymer.

In a third example of this methodology Mei et al. [65] produced a polymer containing the fibronectin-based peptide sequence GRGDS, a well-known cell adhesion sequence. This time, however, the polymerization of hydroxyethyl methacrylate was carried out on the solid support in methyl ethyl ketone/propanol (7 : 3), resulting in a bio-hybrid with a polydispersity of 1.5. This peptide-polymer hybrid material was used as a support for the growth of mouse NIH-3T3 fibroblasts. It was shown that the cells adhered better to the polymer containing the peptide than to the unfunctionalized control polymer. This highlights one of the many potential applications for protein-polymer hybrid materials.

4
Polymer-Protein Conjugates

4.1
PEGylation

One of the most commonly used methods for the creation of a protein-polymer hybrid material is to couple a protein directly to a polymer, using existing functionalities in the protein. One area where these types of protein-polymer hybrid materials have found widespread use is that of the so-called PEGylated therapeutics. It has been shown that by attaching a polyethylene glycol (PEG) chain to a free amine or cysteine within a peptide or protein [70, 71] the physicochemical properties change, shielding the protein or peptide from the body's immune response and increasing the circulation time in the body, allowing the administration of drugs that would otherwise be excreted too rapidly or would not survive in the body [72, 73]. The topic of PEGylation will not be elaborated any further here, because it is well-reviewed in recent literature [1, 74].

4.2
Multi-Site Polymer Conjugation

Another application of peptide-polymer hybrid materials is in the area of tumor targeting. It was noted that the addition of a poly[styrene-co-(maleic anhydride)] chain to the anti-tumor protein neocarzinostatin (Fig. 17), resulted in accumulation of the drug in tumor tissue with a tumor/blood ratio of over 2500, which was much higher than for the protein alone [7]. This was caused by what is known as the enhanced permeability and retention effect (EPR effect) of tumors. This EPR effect is due to the enhanced vascular permeability of tumors when compared to healthy tissue, facilitating the uptake of larger molecules. Therefore, by using protein-polymer hybrid materials, the molecular weight of the agent can be increased, leading to a passive type of targeting [8].

An interesting application of polymer protein conjugates has been reviewed by Alexander et al. [75, 76]. By connecting stimulus responsive synthetic polymers to enzymes, the enzyme activity could be directed by varying temperature or by UV illumination, since this resulted in a change in hydrodynamic volume of the polymers attached [77]. In general, the attachment of polymer chains proved to stabilize the enzymes. In case of the prote-

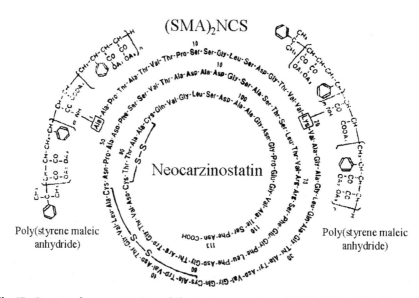

Fig. 17 Structural representation of the anti-tumor agent (SMA)₂NCS, a block copolymer of poly[styrene-co-(maleic anhydride)] coupled to the protein neocarzinostatin. By increasing the molecular weight of the protein, a passive form of tumor targeting occurs due to the increased vascular permeability associated with tumors. Reprinted with permission from [9]. Copyright 2003 Wiley. Adapted from [111]

olytic enzyme trypsin it was demonstrated that the connection of multiple copies of the temperature-sensitive polymer poly(N-ispropyl acrylamide) (poly NIPAAM) could be used for separation of the enzymes from a reaction mixture. Since polyNIPAAM displays LCST behavior, heating the mixture above this LCST temperature resulted in selective precipitation of the biocatalyst. More surprising was the fact that the activity of the enzyme was increased with increasing number of polymer chains attached.

4.3
Site-Directed Polymer Conjugation

Polymer-protein hybrid materials also play an important role in the creation of well-defined, biologically active nano-structured materials, such as vesicles, micelles and rod-like aggregates. There are several examples in which polymers have been coupled to proteins to create interesting supramolecular architectures. Nolte et al. [78] showed it to be possible to synthesize a catalytically active giant amphiphile by coupling the enzyme CAL-B, via a free cysteine group, with maleimide-functionalized polystyrene. Another giant amphiphile was synthesized by Maynard et al. [79] by combining a disulfide-functionalized poly(hydroxy ethyl methacrylate), pHEMA, with bovine serum albumin (BSA).

4.4
Non-Covalent Assemblies

Interestingly, it is not always necessary to connect the protein with the polymer using a covalent bond. It is also possible to use non-covalent methods such as cofactor reconstitution. This was demonstrated by Nolte and co-workers, using two different proteins. In the first example amine-terminated polystyrene was coupled with biotin via a PyBOP coupling. This was then recombined with the protein streptavidin to form giant amphiphiles which, when placed on an air–water interface, assembled into a monolayer. Interestingly, because of the fact that streptavidin has four binding sites the monolayer could be further reacted with a biotinylated ferri-protoporphyrin XI conjugate. Ferri-protoporphyrin XI is a cofactor for horse radish peroxidase (HRP), which after addition to the monolayer created a catalytically active surface [80]. In a second example, ferri-protoporphyrin XI was coupled to amine-terminated polystyrene, again via a standard PyBOP coupling. The cofactor was then reconstituted with HRP to form a giant amphiphile (Fig. 18). The addition of a THF solution of the giant amphiphile to water resulted in the formation of catalytically active vesicles that could be characterized with electron microscopy [81]. Recently, Bontempo and Maynard [82] showed in an elegant alternative approach that streptavidin could be functionalized with synthetic polymers using modified biotin as an initiator for the polymer-

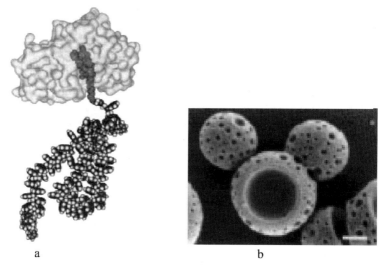

a   b

Fig. 18 **a** Computer-generated model of a HRP-polystyrene-based giant amphiphile; **b** SEM micrograph of an aqueous solution of aggregates formed by the giant amphiphile. Reprinted with permission from [81]. Copyright 2002 Wiley

ization of NIPAAm. They claim the protein is quantitatively modified with polymer and that the resulting polymer is conjugated to the SAv at the biotin binding sites only.

5
Tailor-Made Protein-Polymer Conjugates via Protein Engineering

The above approaches only make use of commercially available proteins. This limits the types of proteins that can be used and the functionalities that are available for polymer modification. To have more control over the sequence and increase the different types of proteins that can be applied researchers have turned to protein engineering. The term "protein engineering" encompasses a whole range of techniques that are focused on altering the expression or properties of proteins in a specific manner. It is applied in a variety of areas such as biochemistry, microbiology, molecular biology and genetics and is used for example to change the cellular localization of enzymes or to modify the properties of enzymes with respect to catalysis and stability. The cloning and expression of genes has become a standard tool in biochemistry and molecular biology [83] and has been recognized only for the last decade as a useful synthetic method for the preparation of polypeptide-based materials.

Protein engineering allows the preparation of monodisperse, high molecular weight polypeptides with complete sequence control. It has been successfully used for the construction of recombinant, structural proteins, such as silks, collagen and elastin [84, 85]. Even block copolymers built up out of two different types of structural proteins such as silk and elastin have been successfully developed [86, 87].

The common feature of these protein polymers is the presence of repetitive sequence motifs which form defined secondary structures. These repetitive amino acid sequences offer the possibility to construct artificial genes by multimerization of small synthetic oligonucleotide sequences and thus the build up of high molecular weight proteins. The constructed artificial genes can be incorporated into an expression plasmid, which can subsequently be transferred to a bacterial host for production of the desired polypeptide (Fig. 19). The most commonly used host is *E. coli*.

The absolute control over the genes that encode for the proteins readily allows the introduction of different functionalities that can be attached to synthetic polymers. Kopeček et al. [88, 89] used this methodology to produce coiled-coil peptide sequences, e.g. $(VSSLESK)_n$ containing a histidine tag on the end. The hydrophobic interaction between the side groups of the valines and leucines in the sequence causes the protein to assume a helical or coil type structure. These hydrophobic interactions then cause several coils to further aggregate to form so called coiled coils. This coiled coil aggregation can be disrupted by temperature, pH and solvent. A copolymer of poly[*N*-(2-hydroxy-propyl) methacrylamide-*co*-(*N'*,*N''*-dicarboxymethylaminopropyl)methacrylamide], (*p*[HPMA-co-DAMA]) with

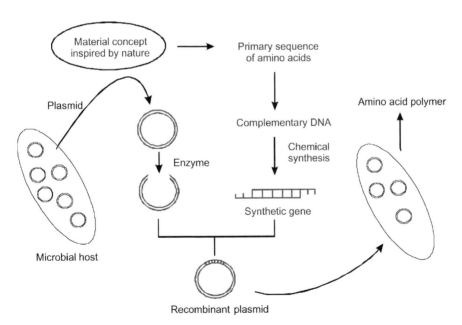

Fig. 19 Overview of the protein engineering methodology [85]

iminotriactetate groups in the side chain was then synthesized using free radical polymerization chemistry. The side groups of this copolymer then formed non-covalent Ni-mediated complexes with the terminal histidine residues of the coiled coil protein. With this approach a series of well-defined, temperature-responsive hydrogels for use in drug delivery applications were produced (Fig. 20).

Another approach was recently developed by Smeenk et al. [90]. They used the combination of protein engineering and polymer modification for the creation of a series of silk-based block copolymers. Spider silk consists of two major domains, a β-sheet crystalline domain, which gives strength to the protein, and a less well-defined amorphous domain, which introduces elasticity and toughness. Smeenk and co-workers produced a β-sheet forming protein

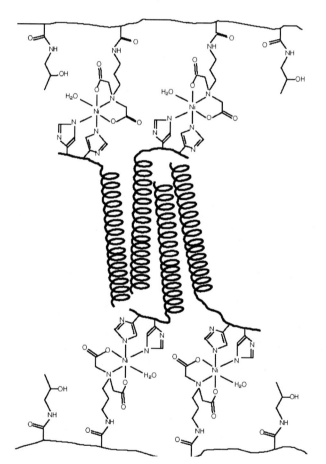

Fig. 20 Structural representation of the hybrid hydrogel of poly(HMPA-co-DAMA) connected to a His-tagged coiled coil via a Ni^{2+} complex. Reprinted with permission from [88]. Copyright 1999 Nature Publishing group

using protein engineering and then coupled this to an amorphous synthetic polymer (poly ethylene oxide). The rationale behind attachment of synthetic polymer blocks at the N- and C-termini was to restrict macroscopic crystallization and to preserve translation of the β-sheet design characteristics of width, height and surface functionality into self-assembled structures. Examination of this hybrid polymer on a surface, using atomic force microscopy (AFM), showed indeed the presence of well-defined fibers.

There are several examples in which protein engineering has been used to introduce specific functionalities at different points in a protein via site-directed mutagenesis, to facilitate the synthesis of peptide polymer hybrid materials. Stayton et al. [91] used protein engineering to create an N49C mutant streptavidin, in which the asparagine at position 49 was substituted

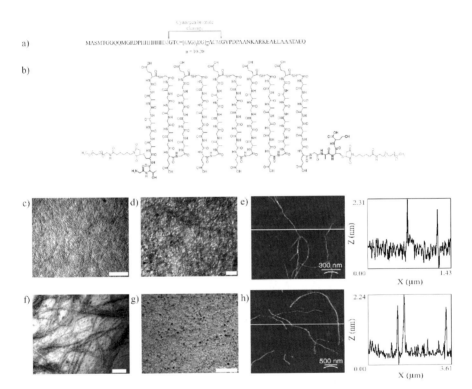

Fig. 21 **a** Sequence of expressed poly-(AG)$_3$EG. **b** ABA-type triblock copolymer prepared by reaction of maleimide-functionalized poly(ethylene glycol)-750 (PEG) with cysteine-flanked poly-[(AG)$_3$EG]$_n$. **c** and **d** Transmission electron micrographs showing fibrillar morphology of crystallized PEG-[(AG)$_3$EG]$_{20}$-PEG conjugate before and after CNBr cleavage, respectively. The scale bars represent 200 nm. **e** Tapping mode atomic force microscopy images of PEG-[(AG)$_3$EG]$_{20}$-PEG fibrils on a mica support. (**f**)–(**g**) as **c**–**d**, but for half the beta-sheet block length (PEG-[(AG)$_3$EG]$_{10}$-PEG). Reprinted with permission from [91]. Copyright 2005 Wiley

for a cysteine. This mutant was then coupled via maleimide chemistry to poly-(N-isopropyl acryl amide) (pNiPAAM). pNiPAAM is known to undergo conformational changes upon heating or changing pH. By attaching pNi-PAAM to streptavidin, Stayton and co-workers were able to create a protein hybrid whose ligand-binding efficiency could be controlled by temperature and pH.

To increase the scope of protein engineering there is broad interest in incorporating non-proteinogenic, or non-natural, amino acids into proteins [85]. By including non-natural amino acids, changes can be made to the protein folding pattern and groups which can facilitate post-translational modification can be introduced. There are several ways to do this. One approach uses the possibility to overwrite stop (nonsense) codons with suppressor tRNAs. If these suppressor tRNAs are functionalized with non-natural amino acids, these artificial building blocks can then become incorporated site specifically into proteins [92]. This method has now been developed to a functional in vivo system, and recently Schultz et al. have shown this approach to be successful to modify a protein with a PEG chain via [2 + 3] Huisgen cycloaddition [93].

An alternative method has been developed by Tirrell et al. [94–96] and involves the substitution of a natural amino acid with a close structural analogue. This method uses bacterial auxotrophs, which are bacteria that have lost the ability to produce one of the natural amino acids. These bacteria are dependent on the medium for their supply of this amino acid and, therefore, if an analogue is added to the medium it can be incorporated instead. The introduction of azido and acetylene analogues of methionine has also led to the presence of bio-orthogonal functional handles that allow further modification of the proteins. However, this methodology has not yet been extended to the incorporation of synthetic polymers.

Protein engineering is an elegant method with which to create peptide-polymer hybrid materials. There are several advantages to using protein engineering, such as complete control over the desired sequence and molecular weight, the absence of polydispersity, and the possibility to synthesize large molecular weight polypeptides and proteins with no deletions or mistakes in the sequence. However, there are several drawbacks which limit the use of protein engineering as a technique for creating materials. Protein engineering is a complex and difficult technique to use, which often requires long periods of time to optimize the production of one peptide sequence, and for every different sequence which is required a new DNA construct has to be synthesized. It is also difficult to produce peptides using protein engineering on a large scale.

The resemblance of structural proteins to segmented multiblock copolymers has instigated researchers to combine blocks from different structural proteins to design new materials, and furthermore get insight into the individual contributions of the blocks on the final material properties. Some examples are given on designed repetitive block copolypeptides that combine motifs like β-sheets in silk, β-spirals in elastin, leucine zipper motifs in DNA-binding proteins and animal cell adhesion sequences in fibronectin.

Multiblock copolymers consisting of silkworm beta-sheet blocks and elastin domains have been reported by Cappello and Ferrari [97]. These silk-elastin-like proteins (SELPs) undergo irreversible solution-to-gel transition under physiological conditions and were investigated as drug delivery systems [98]. Block copolymers with envisioned applications in animal cell culturing and tissue engineered vascular grafts have been reported by several research groups and are based on cell adhesion sequences identified in fibronectin, an extra-cellular multi-adhesive protein that binds to other extracellular matrix components and cell-surface receptors of the integrin family. Cappello and Ferrari reported on silkworm silk-based beta-sheet sequences (GAGAGS)$_n$ containing the cell adhesion RGD motif of fibronectin. The crystalline β-sheet blocks have been described to adsorb to hydrophobic plastic surfaces, while exposing the cell adhesion sequences [97]. Urry et al. and Pan-

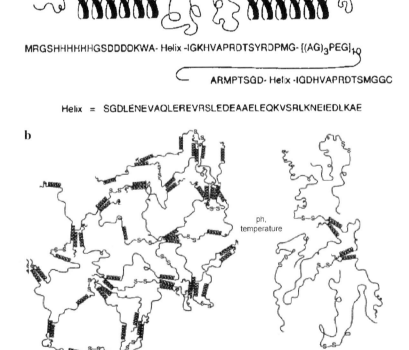

Fig. 22 a Amino acid sequence and b proposed physical gelation of leucine zipper containing triblock copolymer. Reprinted with permission from [101]. Copyright 1998 AAAS

itch et al. used elastin-mimetic sequences in which they either introduced RGD or REDV cell attachment domains, resulting in materials with mechanical properties similar to those of the arterial wall, which supported adhesion of vascular endothelial cells [99, 100].

Coiled-coil motifs are present in a large variety of proteins, like DNA binding proteins, keratins and muscle proteins. An interesting example in which a coiled-coil motif has been used to prepare pH-dependent reversible hydrogels, is the work on triblock copolymers comprising a central random coil, polyelectrolyte domain $[(AG)_3PEG]_{10}$ flanked by leucine zipper domains based on the *Jun* oncogene product [101]. The leucine zipper motif is characterized by a consensus heptad repeat (abcdefg), with a and d the hydrophobic amino acids, whereas e and g are usually charged. The hydrophobic amino acids are located on one side of the helix and cause the formation of a dimeric coiled-coil structure. The charged amino acids modulate the stability of the dimer. The result is a switchable hydrogel: At low pH and low temperature the material formed elastic gels, while gelation was lost upon increase in pH or temperature (Fig. 22).

6
Conclusion and Outlook

The field of polymer-peptide hybrid materials has recently seen a strong increase in activities, catalyzed by the development of new synthetic methodologies that allow a high level of control over the structures to be formed. At each level, whether small peptides or real proteins are concerned, technology has become available to efficiently connect synthetic polymers to peptide moieties. Important in this respect is the ability to build up hybrid materials via a modular approach. The different building blocks, whether they are synthetic polymers or (parts of) proteins can be prepared and extensively characterized prior to their introduction in the final product. This leads to a much better control over the chemical integrity of the hybrid architectures, and enables a better translation of chemical properties into, for example, biological activity. Efficient orthogonal coupling strategies of pre-made building blocks have recently been designed, which will have great impact in the further development of this field in materials science [102]. Besides the native ligation methods [60–63, 103], developed for connecting peptide fragments, also Staudinger ligation [104, 105] and the [3 + 2] dipolar cycloaddition reaction [20, 93, 106–110] have already shown great promise in this respect. Especially in the biomedical field, regarding topics such as targeted drug delivery systems and tissue engineering, it can therefore be expected that in the near future intelligent and creative exploitation of the modular synthetic toolbox for the construction of hybrid materials will lead to improved usage of bioactive moieties.

Acknowledgements The authors would like to acknowledge the Netherlands Technology Foundation (STW) and the Dutch Science Foundation NWO for financial support.

References

1. Duncan R (2003) Nat Rev Drug Disc 2:347
2. Kopeček J (2003) Eur J Pharm Sci 20:1
3. Zuccarello G, Scribner D, Sands R, Buckley LJ (2002) Adv Mater 14:1261
4. Barron AE, Zuckermann RN (1999) Curr Opin Chem Biol 3:681
5. Ratner BD, Bryant SJ (2004) Ann Rev Biomed Eng 6:41
6. Sakiyama-Elbert SE, Hubbell JA (2001) Ann Rev Mater Res 31:183
7. Maeda H, Ueda M, Morinaga T, T M (1984) J Protein Chem 3:181
8. Maeda H (2001) Adv Enzyme Rev 41:189
9. Vandermeulen GWM, Klok HA (2004) Macromol Biosci 4:383
10. Maynard HD, Okada SY, Grubbs RH (2001) Macromolecules 33:6239
11. Jatzkewitz H (1955) Z Naturforsch 10b:27
12. Arshady R (1994) Adv Polym Sci 111:1
13. Batz HG, Franzman G, Ringsdorf H (1972) Angew Chem Int Ed Engl 11:1103
14. Batz HG, Franzman G, Ringsdorf H (1973) Macromol Chem Phys 172:27
15. Arshady R (1984) Macromol Chem Phys 185:2387
16. Arshady R (1981) Macromol Chem Phys 2:573
17. Godwin A, Hartenstein M, Muller AHE, Brocchini S (2001) Angew Chem Int Ed 40:594
18. Schilli CM, Muller AHE, Rizzardo E, Thang SH, Chong YK (2003) ACS Symposium Series 854, p 603
19. Theato P, Kim JU, Lee JC (2004) Macromolecules 37:5475
20. Parrish B, Breitenkamp RB, Emrick T (2005) J Am Chem Soc 127:7404
21. Sanda F, Yokoi M, Kudo H, Endo T (2002) J Polym Sci A1 40:2782
22. Sanda F, Endo T (1999) Macromol Chem Phys 200:2651
23. Murata H, Sanda F, Endo T (2001) Macromol Chem Phys 202:759
24. Murata H, Sanda F, Endo T (1997) Macromol Chem Phys 198:2519
25. Putnam D, Kopeček J (1995) Adv Polym Sci 122:55
26. Kopeček J, Kopeckova P, Minko T, Lu Z-R (2000) J Pharm Biopharm 50:61
27. Perly B, Douy A, Gallot B (1974) CR Acad Sci C Chim 279:1109
28. O'Brien-Simpson NM, Ede NJ, Brown LE, Swan J, Jackson DC (1997) J Am Chem Soc 119:1183
29. Jackson DC, ObrienSimpson N, Ede NJ, Brown LE (1997) Vaccine 15:1697
30. Hopkins TE, Wagener KB (2003) Macromolecules 36:2206
31. Cornelissen JJLM, Donners JJJM, de Gelder R, Graswinckel WS, Metselaar GA, Rowan AE, Sommerdijk NAJM, Nolte RJM (2001) Science 293:676
32. Cornelissen JJLM, Graswinckel SW, Adams PJHM, Nachtegaal GH, Kentgens APM, Sommerdijk NAJM, Nolte RJM (2001) J Polym Sci A1 39:4255
33. Cornelissen J, Fischer M, Sommerdijk N, Nolte RJM (1998) Science 280:1427
34. Vriezema DM, Hoogboom J, Velonia K, Takazawa K, Christianen PCM, Maan JC, Rowan AE, Nolte RJM (2003) Angew Chem Int Ed 42:772
35. Vriezema DM, Kros A, de Gelder R, Cornelissen J, Rowan AE, Nolte RJM (2004) Macromolecules 37:4736
36. Cheuk KKL, Lam JWY, Chen J, Lai LM, Tang BZ (2003) Macromolecules 36:5947

37. Gao GZ, Sanda F, Masuda T (2003) Macromolecules 36:3932
38. Sanda F, Araki H, Masuda T (2004) Macromolecules 37:8510
39. Ayres L, Vos MRJ, Adams PJHM, Shklyarevskiy IO, van Hest JCM (2003) Macro-
    molecules 36:5967
40. Reiersen H, Clarke AR, Rees AR (1998) J Mol Biol 283:255
41. Ayres L, Koch K, Adams PJHM, van Hest JCM (2005) Macromolecules 38:1699
42. Ayres L, Adams HJMP, Löwik WPM, van Hest JCM (2005) Biomacromolecules 6:825
43. Rathore O, Sogah DY (2001) Macromolecules 34:1477
44. Rathore O, Sogah DY (2001) J Am Chem Soc 123:5231
45. Rathore O, Winningham MJ, Sogah DY (2000) J Polym Sci A1 38:352
46. Gosline JM, Demont ME, Denny MW (1986) Endeavour 10:37
47. Yao JM, Xiao D, Chen X, Zhou P, Yu T, Shao Z (2003) Macromolecules 36:7508
48. Bodansky M, Bodansky A (1995) The practice of peptide synthesis (2nd edn).
    Springer Verlag, Berlin
49. Pechar M, Kopeckova P, Joss L, Kopeček J (2002) Macromol Biosci 2:199
50. Vandermeulen GWM, Tziatzios C, Klok HA (2003) Macromolecules 36:4107
51. Rosler A, Klok HA, Hamley IW, Castelletto V, Mykhaylyk OO (2003) Biomacro-
    molecules 4:859
52. Reynhout IC, Lowik DWPM, van Hest JCM, Cornelissen JJLM, Nolte RJM (2005)
    Chem Commun 5:602–604
53. Burkoth TS, Benzinger TLS, Jones DNM, Hallenga K, Meredith SC, Lynn DG (1998)
    J Am Chem Soc 120:7655
54. Burkoth TS, Benzinger TLS, Urban V, Lynn DG, Meredith SC, Thiyagarajan P (1999)
    J Am Chem Soc 121:7429
55. Burkoth TS, Benzinger TLS, Urban V, Morgan DM, Gregory DM, Thiyagarajan P,
    Botto RE, Meredith SC, Lynn DG (2000) J Am Chem Soc 122:7883
56. Eckhardt D, Groenewolt M, Krause E, Börner HG (2005) Chem Commun: 2814
57. Merrifield RB (1997) Solid-Phase Peptide Synthesis (Methods in Enzymology), vol
    289. Elsevier, Amsterdam, p 3
58. Plaue S, Muller S, Briand JP, Vanregenmortel MHV (1990) Biologicals 18:147
59. Canne LE, Botti P, Simon RJ, Chen YJ, Dennis EA, Kent SBH (1999) J Am Chem Soc
    121:8720
60. Botti P, Carrasco MR, Kent SBH (2001) Tetrahedron Lett 42:1831
61. Borgia JA, Fields GB (2000) Trends Biotechnol 18:243
62. Kochendoerfer GG, Chen S, Mao F, Cressman S, Traviglia S, Shao H, Hunter CL,
    Low DW, Cagle EN, Carnevali M, Gueriguian V, Keogh PJ, Porter H, Stratton SM,
    Wiedeke MC, Wilken J, Tang J, Levy JJ, Miranda LP, Crnogorac MM, Kalbag S, Botti P,
    Schindler-Horvat J, Savatski L, Adamson JW, Kung A, Kent SBH, Bradburne JA (2003)
    Science 299:884
63. Rose K, Vizzanova J (1994) J Am Chem Soc 121:30
64. Becker ML, Liu JQ, Wooley KL (2003) Chem Commun: 180
65. Mei Y, Beers KL, Byrd HCM, Vanderhart DL, Washburn NR (2004) J Am Chem Soc
    126:3472
66. Becker ML, Liu J, Wooley KL (2005) Biomacromolecules 6:220
67. He XH, Zhang HL, Wang XY (2002) Polym Bull 48:337
68. Rettig H, Krause E, Borner HG (2004) Macromol Rapid Commun 25:1251
69. Couet J, Jeyaprakash JD, Samuel S, Kopyshev A, Santer S, Biesalski M (2005) Angew
    Chem Int Ed 44:3297
70. Iwai K, Maeda H, Konno T (1984) Cancer Res 44:2115

71. Maeda H, Ueda M, Morinaga TTM (1985) J Med Chem 28:455
72. Bailon P, Berthold W (1998) Pharm Sci Technol Today 1:352
73. Molineux G (2003) Pharmacotherapy 23:3S
74. Duncan R (2002) Brit J Cancer 86:S12
75. de las Heras Alarcon C, Pennadam S, Alexander C (2005) Chem Soc Rev 34:276
76. Cunliffe D, Pennadam S, Alexander C (2004) Eur Polym J 40:5
77. Ding ZL, Chen GH, Hoffman AS (1998) J Biomed Mater Res 39:498
78. Velonia K, Rowan AE, Nolte RJM (2002) J Am Chem Soc 124:4224
79. Bontempo D, Heredia KL, Fish BA, Maynard HD (2004) J Am Chem Soc 126:15372
80. Hannink JM, Cornelissen JJLM, Farrera JA, Foubert P, De Schryver FC, Sommer-dijk NAJM, Nolte RJM (2001) Angew Chem Int Ed 40:4732
81. Boerakker MJ, Hannink JM, Bomans PHH, Frederik PM, Nolte RJM, Meijer EM, Sommerdijk NAJM (2002) Angew Chem Int Ed 41:4239
82. Bontempo D, Maynard HD (2005) J Am Chem Soc 127:6508
83. Sambrook J, Fritsch EF, Maniatis T (1989) Molecular cloning: a laboratory manual. Cold Spring Laboratory Press, Cold Spring
84. Fahnestock SR (2003) Biopolymers: Polyamides and Complex Proteinaceous Materials II. Wiley-VCH, Weinheim
85. van Hest JCM, Tirrell DA (2001) Chem Commun: 1897
86. Dinerman AA, Cappello J, Ghandehari H, Hoag SW (2002) Biomaterials 23:4203
87. Cappello J, Crissman J, Dorman M, Mikolajczak M, Textor G, Marquet M, Ferrari F (1990) Biotechnol Progr 6:198
88. Wang C, Stewart RJ, Kopeček J (1999) Nature 397:417
89. Wang C, Kopeček J, Stewart RJ (2001) Biomacromolecules 2:912
90. Smeenk JM, Otten MBJ, Thies J, Tirrell DA, Stunnenberg HG, van Hest JCM (2005) Angew Chem Int Ed 44:1968
91. Stayton PS, Shimoboji T, Long C, Chilkoti A, Chen GH, Harris JM, Hoffman AS (1995) Nature 378:472
92. Hohsaka T, Kajihara D, Ashizuka Y, Murakami H, Sisido M (1999) J Am Chem Soc 121:34
93. Deiters A, Cropp TA, Summerer D, Mukherji M, Schultz PG (2004) Bioorgan Med Chem Lett 14:5743
94. van Hest JCM, Kiick KL, Tirrell DA (2000) J Am Chem Soc 122:1282
95. Kiick KL, van Hest JCM, Tirrell DA (2000) Angew Chem Int Ed 39:2148
96. Kiick KL, Tirrell DA (2000) Tetrahedron 56:9487
97. Ferrari F, Cappello J (1997) In: McGrath KP, Kaplan DL (eds) Protein-based materials. Birkhauser, Boston, p 37
98. Cappello J, Crissman JW, Crissman M, Ferrari FA, Textor G, Wallis O, Whitledge JR, Zhou X, Burman D, Aukerman L, Stedronsky ER (1998) J Control Release 53:105
99. Nicol A, Gowda DC, Parker TM, Urry DW (1994) Biotechnology of Bioactive Polymers. Plenum Press, New York
100. Panitch A, Yamaoka T, Fournier MJ, Mason TL, Tirrell DA (1999) Macromolecules 32:1701
101. Petka WA, Harden JL, McGrath KP, Wirtz D, Tirrell DA (1998) Science 281:389
102. Yeo DSY, Srinivasan R, Chen GYJ, Yao SQ (2004) Chem-Eur J 10:4664
103. Schwartz EC, Muir TW, Tyszkiewicz AB (2003) Chem Commun: 2087
104. Kiick KL, Saxon E, Tirrell DA, Bertozzi CR (2002) Proc Natl Acad Sci USA 99:19
105. Nilsson BL, Hondal RJ, Soellner MB, Raines RT (2003) J Am Chem Soc 125:5268
106. Dirks AJ, van Berkel SS, Hatzakis NS, Opsteen JA, van Delft FL, Cornelissen JJLM, Rowan AE, Van Hest JCM, Rutjes Fpjt, Nolte RJM (2005) Chem Commun: 4172

107. Opsteen JA, van Hest JCM (2005) Chem Commun: 57
108. Rostovtsev VV, Green LG, Fokin VV, Sharpless KB (2002) Angew Chem Int Ed 41:2596
109. Wang Q, Chan TR, Hilgraf R, Fokin VV, Sharpless KB, Finn MG (2003) J Am Chem Soc 125:3192
110. Wu P, Feldman AK, Nugent AK, Hawker CJ, Scheel A, Voit B, Pyun J, Frechet JMJ, Sharpless KB, Fokin VV (2004) Angew Chem Int Ed 43:3928
111. Duncan R, Dimitrijevic S, Evagorou EG (1996) STP Pharma Sci 6:237
112. Pless J, Boissonnas RA (1963) Helv Chim Act 46:1609
113. Strohriegl P (1993) Macromol Chem Phys 194:363

Adv Polym Sci (2006) 202: 53–73
DOI 10.1007/12_082
© Springer-Verlag Berlin Heidelberg 2006
Published online: 23 February 2006

Solution Properties of Polypeptide-based Copolymers

Helmut Schlaad

Max Planck Institute of Colloids and Interfaces, Colloid Department, Am Mühlenberg 1,
14476 Potsdam-Golm, Germany
schlaad@mpikg-golm.mpg.de

Abstract The aggregation behaviour of biomimetic polypeptide hybrid copolymers and copolypeptides is here reviewed with a particular eye on the occurrence of secondary structure effects. Structure elements like α-helix or β-sheet can induce a deviation from the "classical" phase behaviour and promote the formation of vesicles or hierarchical superstructures with ordering in the length-scale of microns. Polypeptide copolymers are therefore considered as models to study self-assembly processes in biological systems. In addition, they offer a great potential for a production of novel advanced materials and colloids.

Keywords Block copolymer · Peptide · Secondary structure · Self-assembly · Vesicle · Hierarchical structure

Abbreviations
Ac   Acetyl
AUC   Analytical ultracentrifugation
Bzl   Benzyl
CD   Circular dichroism
DIC   Differential interference contrast
DLS   Dynamic light scattering
FTIR   Fourier-transform infrared
LSCM   Laser scanning confocal microscopy
LCST   Lower critical solution temperature
Me   Methyl

NMR   Nuclear magnetic resonance
POM   Polarisation optical microscopy
SANS   Small-angle neutron scattering
SAXS   Small-angle X-ray scattering
SFM   Scanning force microscopy
SLS   Static light scattering
SEM   Scanning electron microscopy
TEM   Transmission electron microscopy
THF   Tetrahydrofuran
TFE   2,2,2-Trifluoroethanol
Z   Benzyloxycarbonyl

1
Introduction

The self-assembly of amphiphilic block copolymers, driven by the incompatibility of constituents, into ordered structures in the sub-micrometre range is a current topic in colloid and materials science [1–5]. Many block copolymers are readily available, and the shape and overall size of the superstructures can be predicted from simple geometric considerations and the lengths of block segments [1]. The basic structures of diblock copolymers in solution are, in the order of decreasing curvature, spherical and cylindrical micelles and vesicles (Fig. 1); the curvature of the core-corona interface is essentially determined by the volume fractions of comonomers and environmental factors (solvent, ionic strength, etc.) [6]. Polymer vesicles, often also referred to as "polymersomes", are particularly interesting as mimetics for biological membranes [7–10].

Deviation from this conventional aggregation behaviour and appearance of more complex superstructures occur, like in biological systems, when specific non-covalent interactions, chirality, and secondary structure effects come into play [11, 12]. Particularly interesting are block copolymers that combine advantageous features of synthetic polymers (solubility, process-

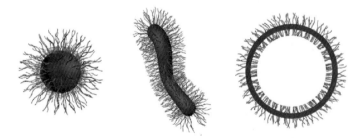

Fig. 1 Illustration of the self-assembled structures of block copolymers in solution: spherical and cylindrical micelle and vesicle (*from left to right*)

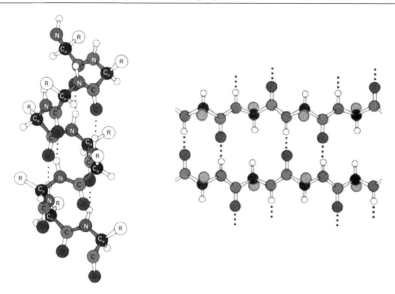

Fig. 2 Illustration of the basic secondary structure motifs of polypeptides: α-helix (*left*) and antiparallel β-sheet (*right*)

ability, rubber elasticity, etc.) with those of polypeptides or polysaccharides (secondary structure, functionality, biocompatibility, etc.). Such biomimetic hybrid polymers or molecular chimeras [13] may produce sophisticated superstructures with new materials properties. "Smart" materials based on polypeptides may reversibly change conformation (α-helix – β-sheet – random coil, Fig. 2) and with it properties in response to an environmental stimulus, like a change in pH or temperature. Also, polypeptide block copolymers may be used as model systems to study generic self-assembly processes in natural proteins.

This review covers the literature on the aggregation of (homo)polypeptide hybrid copolymers and copolypeptides in dilute solution, which was published up to June 2005; a recent review on amphiphiles consisting of peptide sequences is given elsewhere in [12]. It was a particular concern to give a comprehensive overview on secondary structure effects in the self-assembly of these copolymers. Briefly presented are also structures in concentrated solutions (lyotropic phases) and in heterophase systems (see also [14]).

2
Aggregation of Polypeptide Copolymers

Copolymers are classified into peptide hybrid polymers (distinguished further according to the solubility of the polypeptide segment) and copolypeptides. Pronounced effects of the secondary structure on aggregation behaviour should especially be seen when the polypeptide is embedded in the core of aggregates. The ability of polypeptides to crystallise, which is inevitably connected with the α-helical or β-sheet-like secondary structure, however, may be responsible for the appearance of unusual structures out of thermal equilibrium (kinetic control). It is then difficult to draw conclusions on the intrinsic aggregation behaviour of the system. Equilibration is less an issue when the polypeptide is located in the corona of aggregates.

2.1
Polypeptide Hybrid Block Copolymers

2.1.1
Corona-Forming Polypeptides

Block copolymers with soluble and thus corona-forming polypeptide segments include linear diblock and triblock copolymer samples (Fig. 3). In most cases, studies on the aggregation behaviour were done in aqueous solutions with samples consisting of a soft polybutadiene or polyisoprene (exhibiting a glass transition temperature below the freezing point of water) and an α-helical poly(L-glutamate) or poly(L-lysine) segment.

The first study, reported in 1979 by Nakajima et al. [15], dealt with the structure of aggregates of symmetric triblock copolymers consisting of a coiled *trans*-1,4-polybutadiene middle-block and two α-helical poly(γ-benzyl L-glutamate) outer-blocks, $PBLGlu_{53-188}$-b-PB_{64}-b-$PBLGlu_{53-188}$, ("once-broken rods") in chloroform. The shape and dimension of the aggregates in solution were calculated on the basis of simple thermodynamic considerations by taking into account chain conformation and the interfacial free energy. Predictions were found to be in good agreement with the structures (of solvent-cast films) observed by TEM. Depending on the composition of the copolymer, aggregates had a spherical, cylindrical, or lamellar structure with a characteristic size of about 25–45 nm. Similar data were also obtained for block copolymers based on poly(γ-methyl L-glutamate) and poly(Z L-lysine) [16]. These results suggest that the aggregation of rod-coil block copolymers might be treated like that of conformationally isotropic samples [17], provided that the rigid segment is dissolved in the continuous phase (cf. Sect. 2.1.2).

Schlaad et al. [18] and Lecommandoux et al. [19] investigated the aggregates of PB_{27-119}-b-$PLGlu_{20-175}$ in aqueous saline solution by DLS, SLS, SANS, and TEM. Copolymers were found to form spherical micelles with a hydrodynamic radius of $R_h < 40$ nm (70–75 mol % glutamate) or unilamellar vesicles with $R_h = 50$–90 nm (17–54 mol % glutamate); cylindrical micelles have not been observed so far. Against expectations, however, Klok and

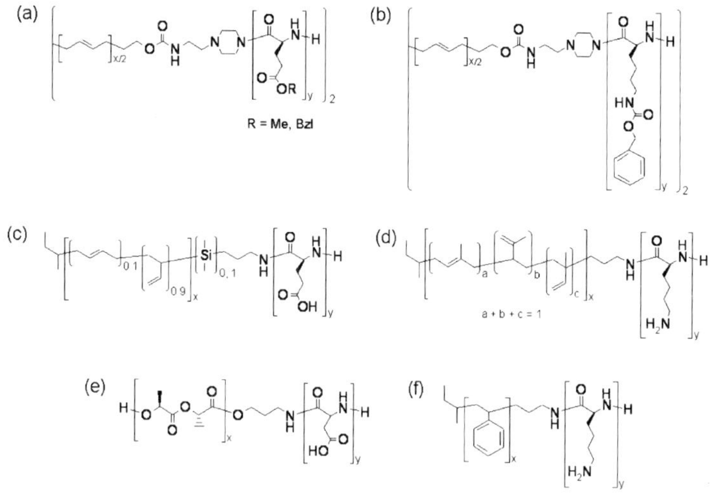

Fig. 3 Chemical structures of linear copolymers (x and y denote number averages of repeating units): **a** poly(γ-methyl L-glutamate)-*block*-1,4-polybutadiene-*block*-poly(γ-methyl L-glutamate) (PMLGlu$_y$-*b*-PB$_x$-*b*-PMLGlu$_y$), poly(γ-benzyl L-glutamate)-*block*-1,4-polybutadiene-*block*-poly(γ-benzyl L-glutamate) (PBLGlu$_y$-*b*-PB$_x$-*b*-PBLGlu$_y$; x = 64, y = 53, 78, 188), **b** poly(Z L-lysine)-*block*-1,4-polybutadiene-*block*-poly(Z L-lysine) (PZLLys$_y$-*b*-PB$_x$-*b*-PZLLys$_y$), **c** 1,2-polybutadiene-*block*-poly(L-glutamic acid) (PB$_x$-*b*-PLGlu$_y$; x = 27–119, y = 20–145), **d** polyisoprene-*block*-poly(L-lysine) (PI$_x$-*b*-PLLys$_y$; x = 49, y = 35–178), **e** poly(L-lactide)-*block*-poly(aspartic acid) (PLL$_x$-*b*-PAsp$_y$; x = 95, 180, y = 47, 70, 270), and **f** polystyrene-*block*-poly(L-lysine) (PS$_x$-*b*-PLLys$_y$; x = 8, 10, y = 9–72)

Lecommandoux et al. [20, 21] earlier reported not micelles but vesicles for a PB$_{40}$-*b*-PLGlu$_{100}$ containing 71 mol % glutamate.

DLS and SANS showed that any pH-induced changes of the secondary structure of poly(L-glutamate) from a random coil (pH > 6) to an α-helix (pH < 5) (CD spectroscopy) did not have severe impact on the morphology (curvature) of the aggregates. SANS further suggested that aggregation numbers remained the same [19], despite equilibration of the sample and a dynamic exchange of polymer chains between aggregates [18]. Coiled and α-helical polypeptide chains seem to have similar spatial requirements at the core-corona interface (see also [15]). However, since the contour length of an all-*trans* polypeptide chain is more than twice that of an α-helix, particularly the hydrodynamic size of the aggregates might decrease when decreasing the pH of the solution. A decrease of the hydrodynamic radius by 20% or less could be observed for PB$_{48}$-*b*-PLGlu$_{56-145}$ [19] but, however, not for PB$_{27-19}$-*b*-PLGlu$_{24-64}$ [18]. PI$_{49}$-*b*-PLLys$_{123}$ micelles in water, also reported by Lecommandoux et al. [22], exhibited a hydrodynamic radius of $R_h \approx 44$ nm at pH 6 (coil) and of 23 nm at pH 11 (helix) (DLS), which corresponds to a decrease of size by almost 50%. Interestingly, although the polypeptide segment is of nearly the same length, the effect seen for PI_{49}-b-$PLLys_{123}$ micelles is much larger than that for PB_{48}-b-$PLGlu_{114}$ ($\sim$ 8%). A possible explanation might be that the poly(L-glutamate) helices are disrupted [16] and/or folded and hence less stretched as poly(L-lysine) helices.

It is noteworthy that the coronae of micelles of PB_{48}-b-$PLGlu_{114}$ and PI_{49}-b-$PLLys_{178}$ could be stabilised using 2,2'-(ethylenedioxy)bisethylamine and glutaric dialdehyde as cross-linking agents, respectively (Lecommandoux et al. [23]). Size and morphology of the aggregates were not affected by the chemical modification reaction.

Ouchi et al. [24] studied the aggregation of $PLL_{95,270}$-b-$PAsp_{47-270}$ in pure water and in 0.2 M phosphate buffer solution at pH 4.4–8.6. Irrespective of the chemical composition of the copolymer (21–74 mol % aspartic acid), however, only spherical aggregates with R_h = 10–80 nm could be observed (DLS/SFM). The aggregation behaviour of the samples was rationalised in terms of a balance between hydrophobic interactions in the poly(L-lactide) core and electrostatic repulsion/hydrogen bridging interactions in the polyaspartate corona. It was not considered at all that the aggregates might be non-equilibrium structures being kinetically trapped in a "frozen" state due to semi-crystallinity of poly(L-lactide) chains. Metastability might be an explanation for the exclusive formation of spherical micelles as well as the seemingly arbitrary changes of the size of aggregates in buffer solutions at different pH (Fig. 4).

Klok et al. [25] used DLS/SLS, SANS, and AUC for the analysis of aggregates formed by $PS_{8,10}$-b-$PLLys_{9-72}$ ($\geq$ 80 mol % lysine) in dilute aqueous solution at neutral pH; at this pH, poly(L-lysine) was in a random coil conformation. They observed, however, cylindrical micelles regardless of the length of the poly(L-lysine) segment. A conclusive explanation of this unexpected aggregation behaviour could not be given. Nonetheless, these results are a first indication that the aggregation of cationic polypeptides could be intrinsically different from that of anionic ones.

Not belonging to the class of polypeptides but worth being mentioned here are and (co)polymers with pendant amino acid residues, that are polyisocyanopeptides (Cornelissen et al. [26–28]) and polyphenylacetylenes carrying L-valine pendants (Bai and Tang et al. [29]). Promoted by the amphiphilic nature and conformational rigidity of the polymer backbone in addition to electrostatic and hydrogen bridging interactions between the amino acids, these polymers were found to self-assemble into a variety of structures including micellar rods and vesicles as well as hierarchical superstructures (filaments, helical cables, etc.).

Complex fibrillar assemblies, among others, were also observed for sodium N-(4-dodecyloxybenzoyl)-L-valinate (Dey et al. [30]) and a lipase polystyrene giant amphiphile (Nolte et al. [31]) in water.

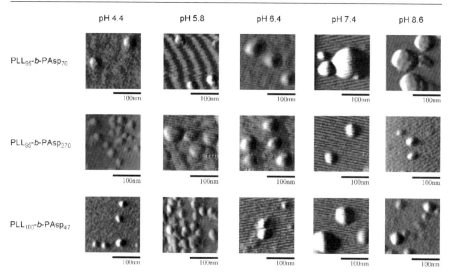

Fig. 4 SFM images of dried aqueous solutions of PLL_x-b-$PAsp_y$ at different pH; preparation of specimens not specified. Reprinted with permission from [24], copyright (2005) American Chemical Society

A [G3]-L-lysine-dendron-substituted cholesteryl-poly(L-lactic acid)$_{22}$ was found to form discrete nano-sized objects ($R_h \sim 15$ nm, DLS) in a dilute aqueous solution (Stupp et al. [32]).

2.1.2
Core-Forming Polypeptides

Aggregates with an insoluble polypeptide core have been prepared with block or random copolymers having linear (Fig. 5) or branched architecture (Fig. 6). Most studies focused on aqueous systems (designated for a use as drug carriers in biomedical applications), and the only inverse systems investigated up to date are PB-b-PLGlu in dilute THF and CH_2Cl_2 solution and PS-b-PZLLys in CCl_4.

Linear Copolymers

Harada and Kataoka [33–35] were the first to investigate the formation of polyion complex (PIC) micelles in an aqueous milieu from a pair of oppositely charged linear polypeptide block copolymers, namely of PEO_{113}-b-$PAsp_{18, 78}$ polyanions and PEO_{113}-b-$PLLys_{18, 78}$ polycations. Complexation studies were carried out at pH 7.29, where both block copolymers had the same degree of ionisation ($\alpha = 0.967$) and were thus double-hydrophilic in nature and not forming aggregates in water. Mixing of the copolymers at a 1 : 1 ratio of amino acid residues resulted in the formation of stable and monodis-

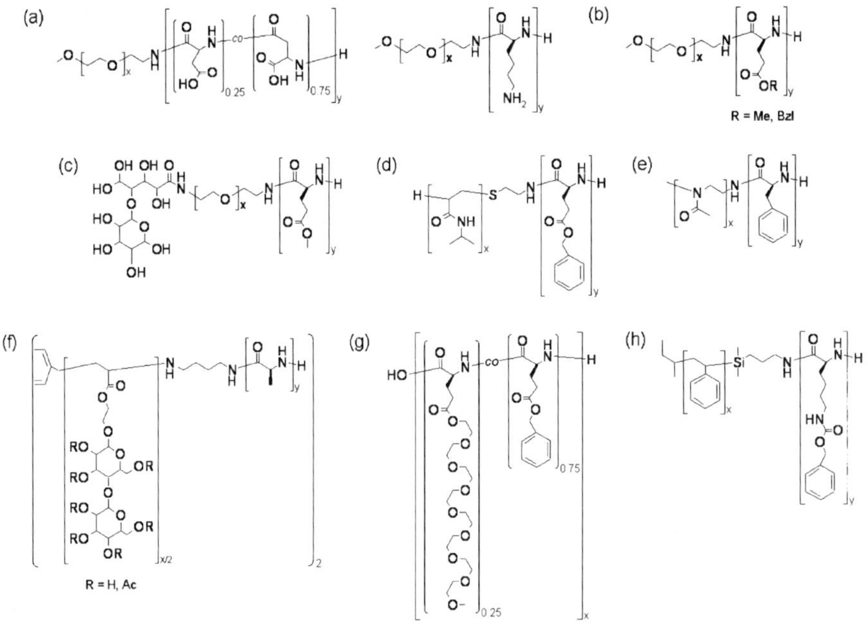

Fig. 5 Chemical structures of linear copolymers (x and y denote number averages of repeating units): **a** poly(ethylene oxide)-*block*-poly(α,β-aspartate) (PEO$_x$-*b*-PAsp$_y$; x = 113, y = 18, 78), poly(ethylene oxide)-*block*-poly(L-lysine) (PEO$_x$-*b*-PLLy$_y$; x = 113, y = 18, 78), **b** poly(ethylene oxide)-*block*-poly(γ-methyl L-glutamate) (PEO$_x$-*b*-PMLGlu$_y$; x = 113, y = 20, 50), poly(ethylene oxide)-*block*-poly(γ-benzyl L-glutamate) (PEO$_x$-*b*-PBLGlu$_y$; x = 272, y = 38, 182, 418), **c** lactose-modified poly(ethylene oxide)-*block*-poly(γ-methyl L-glutamate) (LA-PEO$_x$-*b*-PMLGlu$_y$; x = 75, y = 32), **d** poly(*N*-isopropylacrylamide)-*block*-poly(γ-benzyl L-glutamate) (PNIPAAm$_x$-*b*-PBLGlu$_y$; x = 203, y = 39, 69, 123), **e** poly(*N*-acetyliminoethylene)-*block*-poly(L-phenylalanine) (PAEI$_x$-*b*-PLPhe$_y$; x = 41, y = 4, 8), **f** poly(L-alanine)-*block*-poly(2-acryloylethyl-lactoside)-*block*-poly(L-alanine) (PLAla$_y$-*b*-PAEL$_x$-*b*-PLAla$_y$; x = 9–52, y = 9–93), **g** poly(γ-(ω-methoxy octaoxyethylene) L-glutamate)-*co*-poly(γ-benzyl L-glutamate) ((POLGlu-*co*-PBLGlu)$_x$; x = 775), and **h** polystyrene-*block*-poly(Z L-lysine) (PS$_x$-*b*-PZLLys$_y$; x = 258, y = 57–100)

persed spherical core-shell assemblies of 30 nm in diameter (DLS). Formation of the PIC core is supposedly promoted by the physical (electrostatic) cross-linking of the polypeptide segments rather than by hydrophobic interactions. Due to the stabilizing effect of the poly(ethylene oxide) corona, coalesence of the PIC aggregates to coacervate droplets did not occur like in the homopolymer mixtures of poly(aspartic acid) and polylysine.

Another interesting feature connected with PIC micelles is that of "chain-length recognition" [35]. PIC micelles are exclusively formed by matched pairs of chains with the same block lengths of polyanions and polycations even in mixtures with different block lengths. The key determinants in this recognition process are considered to be the strict phase separation between

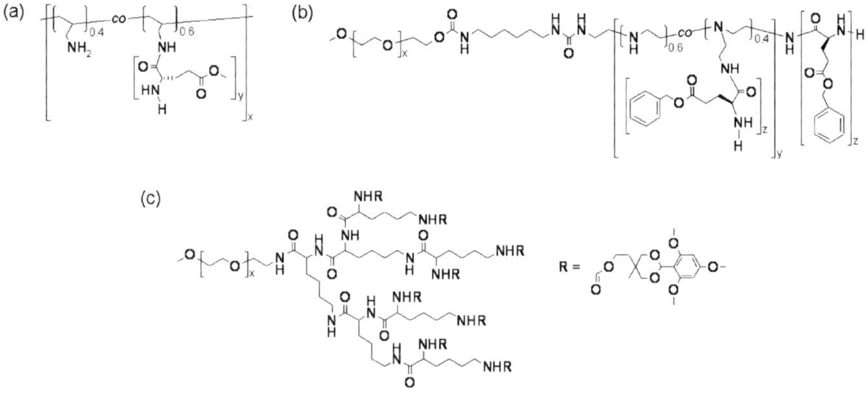

Fig. 6 Chemical structures of branched copolymers (x and y denote number averages of repeating units): **a** polyallylamine-*graft*-poly(γ-methyl L-glutamate) (PAA$_x$-g-PMLGlu$_y$, x = 175, y = 14), **b** poly(ethylene oxide)-*block*-(branched-poly(ethylene imine)-*graft*-poly(γ-benzyl L-glutamate)) (PEO$_x$-b-PEI$_y$-g-PBLGlu$_{z(0.4y+1)}$; x = 113, y = 233, z = 0.2–2.9), and **c** poly(ethylene oxide)-*block*-([G-3]-dendritic poly(L-lysine)-acetal) (PEO$_x$-b-[G-3]-PALLys, x = 113, 227)

the PIC core and the PEO corona, requiring regular alignment of the molecular junctions at the core-corona interface, and the charge stoichiometry (neutralisation). However, referring to own experience, the concept of chain-length recognition does not apply in the complexation of synthetic polyelectrolyte block copolymers.

Yonese et al. [36] studied the aggregation behaviour of PEO$_{113}$-b-PMLGlu$_{20,50}$ and LA-PEO$_{75}$-b-PMLGlu$_{32}$ in water. As shown by DLS, the copolymers formed large aggregates with a hydrodynamic radius of $R_h \approx 250$ nm. Contrary to what was claimed by the authors, it seems more likely that these aggregates were vesicles rather than spherical micelles. Key in the aggregation behaviour might be the association of α-helical poly(γ-methyl L-glutamate) segments, as evidenced by CD spectroscopy, promoting the formation of plane bi-layers which then close into vesicles [5]. Further systematic study on this system and detailed analysis of structures are lacking.

Closely related to this system are PEO$_{272}$-b-PBLGlu$_{38-418}$ and PNIPAAm$_{203}$-b-PBLGlu$_{39-123}$ described by Cho et al. [37, 38]. The aqueous polymer solutions, prepared by the dialysis of organic solutions against water, contained large spherical aggregates ($R_h \approx 250$ nm) with a broad size distribution (DLS). Although the size suggested a vesicular structure of the aggregates, aggregation numbers ($Z < 100$, way of determination not specified) were far below the values of several thousands typically being reported for polymer vesicles [8]. It is also noteworthy that the poly(N-isopropylacrylamide) chains exhibit LCST behaviour. However, raising the temperature to the LCST ($\sim 34\,^\circ$C) had no serious impact on the size of aggregates.

Chaikov and Dong et al. [39, 40] described symmetric triblock copolymers with a glyco methacrylate middle-block and two outer poly(L-alanine) (or poly(γ-benzyl L-glutamate)) blocks. The aggregates formed in dilute aqueous solution were spherical in shape and were 200–700 nm in diameter (TEM). TEM further revealed a compact structure of the aggregates like for multilamellar vesicles. The dimension of the particles, however, was found to decrease with increasing concentration of the copolymer.

Lin et al. [41] investigated the aggregation of a PBLGlu$_{775}$ containing 25 mol % of statistically grafted octa(ethylene glycol) chains in aqueous solution. They observed "regular spindly shaped micelles" with a length of 150–200 nm and a width of about 20 nm (TEM). It is evident, however, that the length of these assemblies is similar to the contour length of the α-helical polypeptide backbone ($l_c \approx 130$ nm). Objects might therefore be nematic bundles of polymer rods, assembled through dipole–dipole interactions, and not micelles. With an estimated diameter of a single polymer rod of 3 nm, a bundle should contain about 40–50 chains.

Naka et al. [42] studied the aggregation behaviour of PAEI$_{41}$-b-PLPhe$_{4, 8}$ in a 0.05 M phosphate buffer at pH 7. Aggregates were observed despite the very low number of hydrophobic L-phenylalanine units, and the size of which was in the order of $R_h = 425$ nm (DLS). The seemingly high tendency of these polymers to form aggregates was attributed to the establishment of hydrogen-bridges between the amino acid units, as shown by IR spectroscopy, in addition to hydrophobic interactions. Visualisation of the aggregates with TEM strongly suggested the presence of coacervates or large clusters of small micelles but no vesicles (Fig. 7). However, the origin of cluster formation has not been discussed in further detail.

The existence of vesicles could be demonstrated for PS$_{258}$-b-PZLLys$_{57}$ in dilute CCl$_4$ solution [43]. SEM imaging (Fig. 8a) showed collapsed hollow spheres of about 300–600 nm in diameter, indicative of vesicles, and also

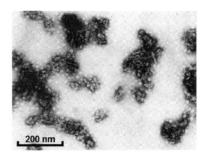

Fig. 7 TEM image of aggregates of PAEI$_{41}$-b-PLPhe$_8$ negatively stained with uranyl acetate; specimen was prepared by deposition of a drop of a 0.2 wt % polymer solution on a carbon-coated copper grid, drawing-off the solution with filter paper, and subsequent drying in vacuo. Reprinted with permission from [42], copyright (1997) Hüthig & Wepf

Fig. 8 SEM images of the aggregates formed by **a** PS_{258}-b-$PZLLys_{57}$ and **b** PS_{258}-b-$PZLLys_{109}$ in dilute CCl_4 solution; specimens were prepared by shock-freezing a 0.2 wt % polymer solution with liquid nitrogen and subsequent freeze-drying [43]

sheet-like structures, supposedly bi-layers that are not yet closed to vesicles [5]. The preference for a lamellar structure might be, like in the previous examples, attributed to a stiffening of the core by the 2D arrangement of crystallisable poly(Z L-lysine) α-helices. PS_{258}-b-$PZLLys_{109}$, on the other hand, was found to form large compact fibrils being hundreds of nanometres in diameter and several tens of microns in length (Fig. 8b); these aggregates might be cylindrical multi-lamellar vesicles. However, the processes involved in the formation of these structures are not known yet.

Likewise, Lecommandoux et al. [19] found vesicles for PB_{48}-b-$PLGlu_{20}$ in THF and in CH_2Cl_2 solution ($R_h = 106$–108 nm, DLS/SLS). The formation of vesicles rather than micelles was attributed to the secondary structure of the insoluble poly(L-glutamate).

Branched Copolymers

PAA_{175}-g-$PMLGlu_{14}$, as reported by Higuchi et al. [44], exhibited a very complex aggregation behaviour in mixtures of water and TFE in dependence of solution composition, pH, and ionic strength (Fig. 9). At pH < 8 and a salt concentration of less than 30 mM, small globular aggregates (micelles) of about 6 nm in diameter were formed, which with time were associating into fibrils of ∼ 4 nm in width and hundreds of nanometres in length (SFM) ("amyloid-like fibril formation"). Simultaneously, and seemingly the driving force of the change in morphology, the poly(γ-methyl L-glutamate) changed conformation from an α-helix to a β-sheet (CD/FTIR). It is noticeable that the complexation of carboxylated poly(ethylene oxide) to the cationic polyallylammonium backbone resulted in stabilisation of the poly(γ-methyl L-glutamate) α-helices and thus inhibition of fibril formation.

Simple geometric considerations suggest that block copolymers with a (hyper-) branched hydrophobic segment of polypeptide grafts or a polypeptide dendron, like the ones described by Chen et al. [45] (PEO_{133}-b-PEI_{233}-g-$PBLGlu_{23-269}$) and Fréchet et al. [46] ($PEO_{113, 227}$-b-[G-3]-PALLys) (Fig. 6),

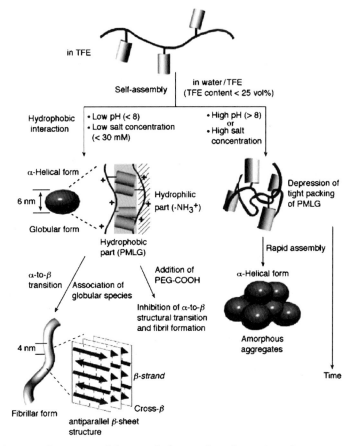

Fig. 9 Schematic illustration of the morphology and conformation of PAA$_{175}$-g-PMLGlu$_{14}$ in water/TFE mixed solvents (abbreviations, PMLG: poly(γ-methyl L-glutamate) and PEG-COOH: ω-carboxy-poly(ethylene glycol)). Reprinted with permission from [44], copyright (2003) Wiley

might give other than spherical micellar aggregates in aqueous solution. However, for these two systems, only spherical micelles with a hydrodynamic diameter of about 170 nm and 25 nm, respectively, could be observed (TEM/DLS). The polypeptide blocks seemed to be too small in size to have a noticeable impact on the curvature of the core-corona interface and thus the morphology of the aggregates. Due to the presence of cationic (PEI) or acidlabile sites (PALLys), micelles were decreasing in size or even disintegrating in an acidic milieu.

2.2
Copolypeptides

Besides the polypeptide hybrid block copolymers described earlier, there exist a few examples of purely peptide-based amphiphiles and block/random copolymers (copolypeptides) (Fig. 10). In the latter case, both the core and corona of aggregates consist of a polypeptide. Any of the studies reported so far dealt with aggregation in aqueous media.

Doi et al. [47, 48] observed the spontaneous formation of aggregates of $PMLGlu_{10}$ with a phosphate head group in water. The freshly prepared solution right after sonication contained globular assemblies (diameter: 50–100 nm, TEM), which after one hour transformed into fibrous aggregates, promoted by intermolecular hydrogen bonding between peptide chains. After one day, these fibrils assembled into a twisted ribbon-like aggregate (TEM). Since its thickness was ~ 4 nm, which is close to the contour length of $PMLGlu_{10}$-P in a β-sheet conformation (CD and FT-IR spectroscopy), the authors concluded that the formation of the ribbon was driven by a stacking of antiparallel β-sheets via hydrophobic interactions (see above) [12].

Lecommandoux et al. [49] showed that zwitterionic $PLGlu_{15}$-b-$PLLys_{15}$ in water can self-assemble into unilamellar vesicles with a hydrodynamic radius of greater than 100 nm (SANS). A change of the pH from 3 to 12 induced an inversion of the structure of the membrane (NMR) and was accompanied by an increase of the size of vesicles from 110 nm to 175 nm (DLS). If the for-

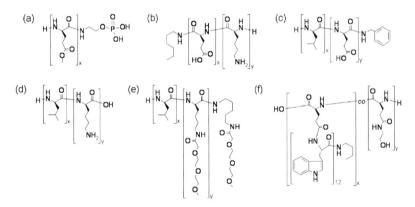

Fig. 10 Chemical structures of peptide-based amphiphiles and copolymers (x and y denote number averages of repeating units): **a** poly(γ-methyl L-glutamate)-phosphate ($PMLGlu_x$-P; x = 10), **b** poly(L-glutamic acid)-*block*-poly(L-lysine) ($PLGlu_x$-b-$PLLys_y$; x = y = 15), **c** poly(L-leucine)-*block*-poly(L-glutamic acid) ($PLLeu_x$-b-$PLGlu_y$; x = y = 180), **d** poly(L-leucine)-*block*-poly(L-lysine) ($PLLeu_x$-b-$PLLys_y$; x = 10–40, y = 90—380), **e** poly(L-leucine)-*block*-poly(N^ε-[2-(2-(methoxyethoxy)ethoxy)]acetyl L-lysine) ($PLLeu_x$-b-$PELLys_y$; x = 10–75, y = 60–200), and **f** poly(N-hydroxyethyl L-glutamine)-*co*-poly(γ-poly(L-tryptophan) L-glutamine) (($PHLGln$-co-$P(PLTrp)LGln)_x$, x = 81–367, x + y = 1400)

mation of vesicles was controlled by a secondary structure effect or simply by copolymer composition (geometry) remains an open question. Spectroscopic data supporting an α-helical conformation of the polypeptide in the hydrophobic part of the membrane, as speculated by the authors, were not provided. Also, the peptide segments seem to be too short to form a stable α-helix [50].

Meyrueix et al. [51] performed the selective precipitation of $PLLeu_{180}$-b-$PLGlu_{180}$ and obtained nanoparticles, which could be purified and further suspended in water or in a 0.15 M phosphate saline buffer at pH 7.4. The colloidal dispersions were stable, due to the electrosteric stabilisation of the particles by poly(sodium L-glutamate) brushes, containing spherical or cylindrical micelles besides large hexagonally-shaped platelets with a diameter of about 200 nm (TEM, Fig. 11). The different shapes of particles were due to the heterogeneity of copolymer chains with respect to chemical composition (NMR): glutamate-rich chains formed micelles and leucine-rich ones formed platelets. CD spectroscopy and X-ray diffraction suggested that the core of platelets consisted of crystalline, helical poly(L-leucine) segments, and the structural driving force was thus related to the formation of leucine zippers in a 3D array.

Deming et al. [52] introduced non-ionic block copolypeptides of L-leucine and ethylene glycol-modified L-lysine residues, $PLLeu_{10-75}$-b-$PELLys_{60-200}$. The copolymers adopted a rod-like conformation, due to the strong tendency of both segments to form α-helices, as confirmed by CD spectroscopy. The self-assembled structures observed in aqueous solutions included (sub-)micrometre vesicles, sheet-like membranes, and irregular aggregates (Fig. 12). Exemplarily, upon dissolution of $PLLeu_{20}$-b-$PELLys_{100}$ at a concentration of 0.5–3.0 wt % in water, micron-sized spherical assemblies in addition to a few sheet-like and irregular structures were formed, as revealed by DIC optical microscopy. When the copolymer sample was first dissolved in

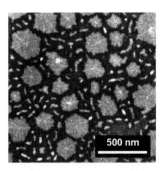

Fig. 11 TEM image of particles of $PLLeu_{180}$-b-$PLGlu_{180}$ (sodium salt) negatively stained with silicotungstate; specimen was prepared by deposition of a drop of a 5.6 wt % polymer solution in water on a carbon-coated copper grid and subsequent removal of the excess of solution. Reprinted with permission from [51], copyright (1999) Academic Press

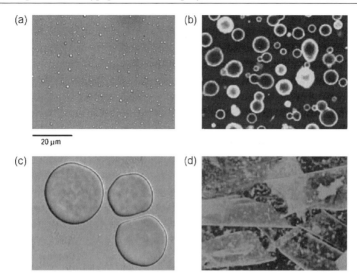

Fig. 12 Images of different PLLeu$_x$-b-PELLys$_y$ samples for 1.0 wt% suspensions in de-ionised water: DIC optical micrographs of **a** PLLeu$_{20}$-b-PELLys$_{60}$ and **c** PLLeu$_{40}$-b-PELLys$_{150}$ samples and LSCM images of a suspension of **b** PLLeu$_{20}$-b-PELLys$_{100}$ and **d** PLLeu$_{40}$-b-PELLys$_{200}$ visualised with the fluorescent DiOC$_{18}$ dye. Reprinted with permission from [52], copyright (2004) Macmillan Publishers Ltd

an organic solvent and then dialysed against water, only spherical assemblies of 5–10 μm in diameter were observed, which were identified as unilamellar giant vesicles (LSCM, Fig. 12b, and SANS). The assembly into bi-layers, which is an atypical phenomenon for non-ionic block copolymers of this composition, was related to a secondary structure effect. Accordingly, racemic samples with the same composition but non-helical chain conformation (CD) were not found to form giant vesicles.

It is further noticeable that vesicles gained pH-responsiveness when L-leucine residues in the hydrophobic part were statistically substituted by L-lysine. Acidification of the solution resulted in protonation of amine functionalities and the near-instantaneous disruption of the membrane of vesicles.

Deming et al. [53–55] also described the formation of hydrogels from di- and triblock copolypeptides based on PLLeu$_{10-40}$ and PLLys$_{90-380}$ (hydrobromide); gelation occurred at a polymer concentration as low as 0.25 wt%. It is thought that the scaffold of the hydrogel is built of twisted fibril assemblies with a core of closely packed poly(L-leucine) α-helices and a corona of cationic poly(L-lysine) chains.

Okamoto et al. [56] analysed the aqueous solutions of branched (PHLGln-co-P(PLTrp)$_{12}$LGln)$_{1400}$ consisting of a poly(N-hydroxyethyl L-glutamine) backbone and varying amounts of short, statistically grafted poly(L-tryptophan) chains (6–26 mol% tryptophan). Stable spherical aggregates with a diameter of about 80 nm (DLS) were observed when the amount of hy-drophobic residues was in the range of 14–23 mol %. TEM suggested that the aggregates did not have a core-shell structure. Instead, the poly(L-tryptophan) chains were forming small hydrophobic compartments of about 5 nm in diameter (SAXS: 4.5 nm), which were statistically spread in a matrix of the hydrophilic poly(N-hydroxyethyl L-glutamine). The occurrence of this unusual morphology was explained in terms of greater loss of entropy and larger interface area of graft copolymers in comparison to a linear diblock copolymer system.

3
Lyotropic Phases of Polypeptide Copolymers

Polypeptides in an α-helical conformation can act as rod-like mesogens, and thus polypeptide block copolymers exhibit a high tendency towards the for-mation of lamellar phases in concentrated solutions (i.e. lyotropic phases) and in the bulk. PS-b-PBLGlu, for instance, forms a hexagonal-in-lamellar structure in the bulk (i.e. alternating sheets of polystyrene and poly(γ-benzyl L-glutamate) with the helices being arranged in a 2D hexagonal array) with a lamellar spacing of some tens of nanometres [14].

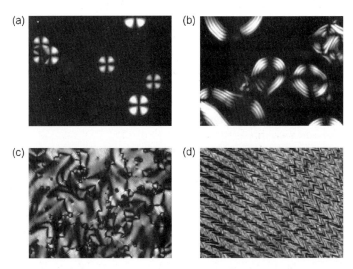

Fig. 13 POM images of textures observed under crossed polarisers for $\sim$ 20 wt % organic solutions of PS$_x$-b-PBLGlu$_y$: **a** PS$_{52}$-b-PBLGlu$_{29}$ in 1,4-dioxane, lamellar phase, **b** PS$_{218}$-b-PBLGlu$_{598}$ in 1,4-dioxane, twisted cholesteric phase, **c** PS$_{52}$-b-PBLGlu$_{90}$ in 1,4-dioxane (solid film), cholesteric phase, and **d** PS$_{40}$-b-PBLGlu$_{80}$ in chloroform (solid film), hex-agonal phase [43]

Gallot et al. [57] investigated the swelling behaviour of PS-*b*-PBLGlu films in a good solvent for either component, namely 1,4-dioxane. SAXS analyses showed that the films maintained their morphology but with a 50% larger spacing of lamellae. Results further indicated that the increase in thickness was solely due to the swelling of the polystyrene layers; the thickness of the (crystalline) poly(γ-benzyl L-glutamate) layers, on the other hand, remained unaffected by the solvent. The inverse situation was observed for PB-*b*-PLLys and water as the swelling agent, that is the polypeptide layers expanded and the organic layers maintained thickness [58].

Additional structural order in the lyotropic phases and bulk films of PBLGlu-based block copolymers beyond the nanometre-length scale could be revealed by POM (Fig. 13, Łosik [43]) and TEM/SFM (Pochan et al. [59]). Smectic, hexagonal, (twisted) cholesteric, and other phases could be observed depending on the chemical composition of the copolymer and the history and processing of the sample. A comprehensive picture of the processes involved in the formation of hierarchical superstructures does not exist yet.

4
Stabilizing Properties of Polypeptide Copolymers

Block copolymers with polypeptide segments were occasionally used to stabilise oil-in-water emulsions or as emulsifiers in heterophase polymerisation processes.

Gallot et al. [60] used lipopeptides to produce stable water-in-oil macroemulsions with a droplet diameter greater than 1 μm. In the presence of a cosurfactant, these emulsions could be transformed into miniemulsions containing droplets of 100–400 nm in diameter (cetyl alcohol) or microemulsions with droplets of 10–100 nm (C4 amine or alcohol). Lipopeptide-alkyl(meth)acrylamide copolymers and acrylamidoliposarcosine-alkylacrylamide copolymers were further used as emulsifiers of cosmetic oils [61].

Kimizuka et al. [62] obtained microemulsions upon mixing PBLGlu$_{12, 32}$-COOH in CH_2Cl_2 with an aqueous NaOH solution. Evaporation of the organic solvent resulted in the formation of hollow spheres (microcapsules) being 1–7 μm in diameter.

Schlaad et al. [63] used polystyrene-poly(sodium glutamate) block copolymers with linear and multi-arm star architectures to produce electrosterically stabilised polystyrene latexes with a polypeptide decoration (diameter: 70–220 nm). The main latex properties, that are the average particle size and distribution, electrolyte stability and electrophoretic mobility, were strongly affected by the architecture of the stabiliser. The star-shaped stabilisers exhibit a higher stabilizing efficiency than the linear analogues and are able to stabilise a second generation of particles producing bimodal latex dispersions

(Fig. 14). The latexes were stable in physiological salt solutions, which is an important feature for any medicinal application.

Also worth mentioning is the application of polypeptide-based copolymers in the production of magnetic nanocomposite materials.

Lecommandoux et al. [64] obtained stable dispersions of super-paramagnetic micelles and vesicles by combining aqueous solution of PB_{48}-b-$PLGlu_{56-145}$ with a ferrofluid consisting of maghemite (γ-Fe_2O_3) nanoparticles in CH_2Cl_2. Incorporation of one mass equivalent of ferrofluid into the hydrophobic core of aggregates did not alter their morphology, as deducted from SLS and SANS data, but caused a substantial increase of the outer diameter by a factor of 6 (DLS). Interestingly, the hybrid vesicles underwent deformation under a magnetic field, as shown by 2D-SANS experiments.

Held et al. [65] earlier reported that monodisperse, highly crystalline maghemite nanoparticles in organic solvents could be transferred into an aqueous medium using tetramethylammonium hydroxide and stabilised at

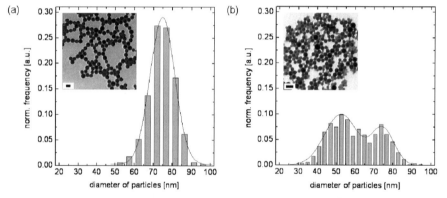

Fig. 14 Particle size distributions of latexes obtained with **a** a linear PS_{52}-b-$PNaGlu_{104}$ and **b** a star-shaped PS_{52}-b-$(PNaGlu_{22})_8$. *Insets* show the corresponding TEM images (scale bar = 100 nm) of dried latex samples [63]

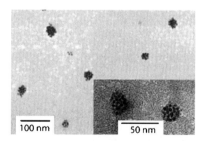

Fig. 15 TEM images of clusters of maghemite nanoparticles deposited from dispersions in water in the presence of $PELLys_{100}$-b-$PAsp_{30}$. Reprinted with permission from [65], copyright (2003) American Chemical Society

neutral pH. Combination of the aqueous maghemite solution with $PELLys_{100}$-b-$PAsp_{30}$ led to the formation of uniform clusters comprising approximately 20 nanoparticles (Fig. 15).

5
Summary

The phase behaviour of biomimetic polypeptide-based copolymers in solution was described and discussed with respect to the occurrence of secondary structure effects. Evidently, incorporation of crystallisable polypeptide segments inside the core of an aggregate has impact on the curvature of the core-corona interface and promotes the formation of fibrils or vesicles or other flat superstructures. Spherical micelles are usually not observed. Copolymers with soluble polypeptide segments, on the other hand, seem to behave like conventional block copolymers. A pH-induced change of the conformation of coronal polypeptide chains only affects the size of aggregates but not their shape. The lyotropic phases of polypeptide copolymers indicate the existence of hierarchical superstructures with ordering in the length-scale of microns.

It is evident that the biomimetic approach using polypeptide hybrid polymers is very successful in the creation of novel superstructures with hierarchical order. Although started about 30 years ago in the mid-1970s, this field is still in a premature state. The greatest number of systematic studies on the aggregation behaviour in solution is younger than 5 to 10 years. A comprehensive picture of the processes involved in the formation of hierarchical superstructures has not yet emerged.

The application potential of polypeptide copolymers has also not been exhausted. Most studies deal with ordinary micelles for the controlled delivery of drugs or genes. Not much attention has, for whatever reason, been paid to other colloidal systems like for instance emulsions, polymer latexes, and inorganic–organic biohybrid nanoparticles.

Acknowledgements The author would like to thank his former coworkers Hildegard Kukula, Magdalena Łosik, and Rémi Soula for working in the field of polypeptide hybrid polymers and also Markus Antonietti and Erich C. Financial support was given by the Max Planck Society.

References

1. Förster S, Antonietti M (1998) Adv Mater 10:195
2. Bates FS, Fredrickson GH (1999) Physics Today 52:32
3. Cölfen H (2001) Macromol Rapid Commun 22:219
4. Förster S, Konrad M (2003) J Mater Chem 13:2671
5. Antonietti M, Förster S (2003) Adv Mater 15:1323

6. Choucair A, Eisenberg A (2003) Eur Phys J E 10:37
7. Discher BM, Won Y-Y, Ege DS, Lee JC-M, Bates FS, Discher DE, Hammer DA (1999) Science 284:1143
8. Discher DE, Eisenberg A (2002) Science 297:967
9. Taubert A, Napoli A, Meier W (2004) Curr Opin Chem Biol 8:598
10. Kita-Tokarczyk K, Grumelard J, Haefele T, Meier W (2005) Polymer 46:3540
11. Cornelissen JJLM, Rowan AE, Nolte RJM, Sommerdijk NAJM (2001) Chem Rev 101:4039
12. Löwik DWPM, van Hest JCM (2004) Chem Soc Rev 33:234
13. Schlaad H, Antonietti M (2003) Eur Phys J E 10:17
14. Gallot B (1996) Prog Polym Sci 21:1035
15. Nakajima A, Kugo K, Hayashi T (1979) Macromolecules 12:844
16. Hayashi T (1985) In: Goodman I (ed) Developments in Block Copolymers. Elsevier, London, p 109
17. Förster S, Zisenis M, Wenz E, Antonietti M (1996) J Chem Phys 104:9956
18. Kukula H, Schlaad H, Antonietti M, Förster S (2002) J Am Chem Soc 124:1658
19. Chécot F, Brûlet A, Oberdisse J, Gnanou Y, Mondain-Monval O, Lecommandoux S (2005) Langmuir 21:4308
20. Chécot F, Lecommandoux S, Gnanou Y, Klok H-A (2002) Angew Chem Int Ed Engl 41:1340
21. Chécot F, Lecommandoux S, Klok H-A, Gnanou Y (2003) Eur Phys J E 10:25
22. Babin J, Rodríguez-Hernández J, Lecommandoux S, Klok H-A, Achard M-F (2005) Faraday Discuss 128:179
23. Rodríguez-Hernández J, Babin J, Zappone B, Lecommandoux S (2005) Biomacromolecules 6:2213
24. Arimura H, Ohya Y, Ouchi T (2005) Biomacromolecules 6:720
25. Lübbert A, Castelletto V, Hamley IW, Nuhn H, Scholl M, Bourdillon L, Wandrey C, Klok H-A (2005) Langmuir 21:6582
26. Cornelissen JJLM, Fischer M, Sommerdijk NAJM, Nolte RJM (1998) Science 280:1427
27. Cornelissen JJLM, Fischer M, van Waes R, van Heerbeek R, Kamer PCJ, Reek JNH, Sommerdijk NAJM, Nolte RJM (2004) Polymer 45:7417
28. Cornelissen JJLM, Donners JJJM, de Gelder R, Graswinckel WS, Metselaar GA, Rowan AE, Sommerdijk NAJM, Nolte RJM (2001) Science 293:676
29. Li BS, Cheuk KKL, Ling LS, Chen JW, Xiao XD, Bai CL, Tang BZ (2003) Macromolecules 36:77
30. Mohanty A, Dey J (2004) Langmuir 20:8452
31. Velona K, Rowan AE, Nolte RJM (2002) J Am Chem Soc 124:4224
32. Klok H-A, Hwang JJ, Hartgerink JD, Stupp SI (2002) Macromolecules 35:6101
33. Harada A, Kataoka K (1995) Macromolecules 28:5294
34. Harada A, Kataoka K (1997) J Macromol Sci Pure Appl Chem A34:2119
35. Harada A, Kataoka K (1999) Science 283:65
36. Toyotama A, Kugimiya S-I, Yamanaka J, Yonese M (2001) Chem Pharm Bull 49:169
37. Cheon J-B, Jeong Y-I, Cho C-S (1998) Korea Polym J 6:34
38. Cheon J-B, Jeong Y-I, Cho C-S (1999) Polymer 40:2041
39. Dong C-M, Sun X-L, Faucher KM, Apkarian RP, Chaikof EL (2004) Biomacromolecules 5:224
40. Dong C-M, Faucher KM, Chaikof EL (2004) J Polym Sci Polym Chem 42:5754
41. Tang D, Lin J, Lin S, Zhang S, Chen T, Tian X (2004) Macromol Rapid Commun 25:1241

42. Naka K, Yamashita R, Nakamura T, Ohki A, Maeda S (1997) Macromol Chem Phys 198:89
43. Losik M (2004) Dissertation, University of Potsdam, Germany
44. Koga T, Taguchi K, Kobuke Y, Kinoshita T, Higuchi M (2003) Chem Eur J 9:1146
45. Tian HY, Deng C, Lin H, Sun J, Deng M, Chen X, Jing X (2005) Biomaterials 26:4209
46. Gillies ER, Jonsson TB, Fréchet JMJ (2004) J Am Chem Soc 126:11936
47. Doi T, Kinoshita T, Kamiya H, Tsujita Y, Yoshimizu H (2000) Chem Lett 262
48. Doi T, Kinoshita T, Kamiya H, Washizu S, Tsujita Y, Yoshimizu H (2001) Polym J 33:160
49. Rodríguez-Hernández J, Lecommandoux S (2005) J Am Chem Soc 127:2026
50. Harada A, Cammas S, Kataoka K (1996) Macromolecules 29:6183
51. Constancis A, Meyrueix R, Bryson N, Huille S, Grosselin J-M, Gulik-Krzywicki T, Soula G (1999) J Colloid Interface Sci 217:357
52. Bellomo EG, Wyrsta MD, Pakstis L, Pochan DJ, Deming TJ (2004) Nat Mater 3:244
53. Nowak AP, Breedveld V, Pakstis L, Ozbas B, Pine DJ, Pochan D, Deming TJ (2002) Nature 417:424
54. Pochan DJ, Pakstis L, Ozbas B, Nowak AP, Deming TJ (2002) Macromolecules 35:5358
55. Breedveld V, Nowak AP, Sato J, Deming TJ, Pine DJ (2004) Macromolecules 37:3943
56. Okamoto S, Hanai T, Yamauchi F, Nagata K, Sakurai S, Sakurai K, Nakanishi E (2001) Mater Sci Res Int 7:219
57. Douy A, Gallot B (1982) Polymer 23:1039
58. Billot J-P, Douy A, Gallot B (1976) Makromol Chem 177:1889
59. Minich EA, Nowak AP, Deming TJ, Pochan DJ (2004) Polymer 45:1951
60. Gallot B, Hassan HH (1989) ACS Symposium Series 384:116
61. Gallot B (1991) ACS Symposium Series 448:103
62. Morikawa M-A, Yoshihara M, Endo T, Kimizuka N (2005) Chem Eur J 11:1574
63. Kukula H, Schlaad H, Tauer K (2002) Macromolecules 35:2538
64. Lecommandoux S, Sandre O, Chécot F, Rodríguez-Hernández J, Perzynski R (2005) Adv Mater 17:712
65. Euliss LE, Grancharov SG, O'Brien S, Deming TJ, Stucky GD, Murray CB, Held GA (2003) Nano Lett 3:1489

Adv Polym Sci (2006) 202: 75–111
DOI 10.1007/12_083
© Springer-Verlag Berlin Heidelberg 2006
Published online: 23 March 2006

Solid-State Structure, Organization and Properties of Peptide— Synthetic Hybrid Block Copolymers

Harm-Anton Klok[1] (✉) · Sébastien Lecommandoux[2] (✉)

[1]Ecole Polytechnique Fédérale de Lausanne (EPFL), Institut des Matériaux, Laboratoire des Polymères, STI – IMX – LP, MXD 112 (Bâtiment MXD), Station 12, 1015 Lausanne, Switzerland
harm-anton.klok@epfl.ch

[2]Laboratoire de Chimie des Polymères Organiques, LCPO (UMR5629) – CNRS – ENSCPB, Université Bordeaux 1, 16 Avenue Pey Berland, 33607 Pessac cedex, France
lecommandoux@enscpb.fr

Abstract Peptide–synthetic hybrid block copolymers are most easily prepared via ring-opening polymerization of α-amino acid N-carboxyanhydrides using appropriately end-functionalized synthetic polymers as the macroinitiator. This class of peptide–synthetic hybrid block copolymers has been the subject of longstanding scientific interest. First reports on the structure and properties of these hybrid block copolymers date from the 1970s. The advent of more refined analytical tools has spurred the interest in these materials and has led to an increased understanding of their structure and properties. This article presents an overview of the solid-state structure, organization and properties of the major classes of peptide–synthetic hybrid block copolymers, classified according to the chemical composition of the synthetic polymer block. For each of these classes of block copolymers, results from earlier morphological studies will be discussed and compared and complemented with more recent results. Properties and possible applications of peptide–synthetic hybrid block copolymers will also be pointed out.

Keywords Block copolymers · Polypeptides · Self-assembly

Abbreviations

AFM	atomic force microscopy
BF	bovine fibrinogen
BγG	bovine γ globulin
BPF	bovine plasma fibrinogen
BSA	bovine serum albumine
CD	circular dichroism
DMF	N,N-dimethylformamide
DSC	differential scanning calorimetry
FDA	U.S. Food and Drug Administration
FTIR	Fourier transform infrared
IR	infrared
PB	polybutadiene
PBA	poly(butadiene-co-acrylonitrile)
PBLA	poly(β-benzyl-L-aspartate)
PBDLG	poly(γ-benzyl-D,L-glutamate)
PBLG	poly(γ-benzyl-L-glutamate)
PCL	poly(ε-caprolactone)
PDMS	polydimethylsiloxane
PEG	poly(ethylene glycol)
PECF	pseudoextracellular fluid
PELG	poly(γ-ethyl-L-glutamate)
PEUU	polyetherurethaneurea
PHF	poly(9,9-dihexylfluorene-2,7-diyl)
PHPG	poly(N^5-hydroxypropyl-L-glutamine)
PI	polyisoprene
PLA	poly(L-alanine)
PLL	poly(L-lysine)
PLLA	poly(L-lactic acid)
PMLG	poly(γ-methyl-L-glutamate)
PMDLG	poly(γ-methyl-D,L-glutamate)
PPG	poly(propylene glycol)
PS	polystyrene
PTHF	poly(tetrahydrofuran)
PZLL	poly(ε-benzyloxycarbonyl-L-lysine)
SAXS	small angle X-ray scattering
TFA	trifluoroacetic acid
T_g	glass transition temperature
TEM	transmission electron microscopy
THF	tetrahydrofuran
WAXS	wide angle X-ray scattering
XPS	X-ray photoelectron spectroscopy

1
Introduction

Now that the fundamental principles that underlie the self-assembly of block copolymers have been addressed in numerous theoretical and experimental studies, these materials are finding increasing interest in several nanotechnology applications such as nanostructured membranes, templates for nanoparticle synthesis and high density information storage [1–3]. The self-assembly of block copolymers composed of two (or more) chemically incompatible, amorphous segments is determined by the interplay of two competitive processes [4, 5]. On the one hand, in order to avoid unfavourable monomer contacts, the blocks segregate and try to minimize interfacial area. Minimization of interfacial area, however, involves chain stretching, which is entropically unfavourable. It is the interplay between these two processes that determines the final block copolymer morphology. Depending on the volume fractions of the respective blocks, amorphous AB diblock copolymers can form lamellar, hexagonal, spherical and gyroid structures. More complex morphologies can also be generated, but these require alternative strategies. One possibility is to increase the number of different blocks [6]. The phase behaviour of a linear ABC triblock copolymer, for example, is described by three binary interaction parameters χ, two independent composition variables and three possible sequences. In contrast, the phase behaviour of an AB diblock copolymer can be described by only two variables; one binary interaction parameter and one composition variable. Other strategies to create more complex block copolymer nanostructures include amongst others the introduction of more rigid or (liquid) crystalline blocks [7–9] and the complexation of low molar mass amphiphiles to polyelectrolyte block copolymers [10, 11]. Often, such conformationally restricted segments introduce additional secondary interactions such as electrostatics, hydrogen bonding or $\pi - \pi$ interactions that also influence block copolymer self-assembly. In particular, rod-coil type block copolymers composed of a rigid (crystalline) block and a flexible (amorphous) segment have attracted increased interest over the past years [12, 13]. Morphological studies on rod-coil block copolymers have revealed several unconventional nanoscale structures, which were not previously known for purely amorphous block copolymers. These findings underline the potential of manipulating chain conformation and interchain interactions to further engineer block copolymer self-assembly.

This review discusses the solid-state structure, organization and properties of peptide–synthetic hybrid block copolymers. Hybrid block copolymers composed of a synthetic block and a peptide segment are an interesting class of materials, both from a structural and a functional point of view [14–16]. Peptide sequences can adopt ordered conformations such as α-helices or β-strands. In the former case, this leads to block copolymers with rod-coil character. Peptide sequences with a β-strand conformation can undergo intermolecular hydrogen-bonding, which also offers additional means to direct nanoscale structure formation compared to purely amorphous block copolymers. Combining peptide sequences and synthetic polymers, however, is not only interesting to enhance control over nanoscale structure formation, but can also result in materials that can interface with biology. Obviously, such materials are of great potential interest for a variety of biomedical and bioanalytical applications.

The interest in peptide–synthetic hybrid block copolymers has considerably increased over the past years and there is a vast amount of literature describing the synthesis, organization and properties of a wide range of different block copolymers. This article will exclusively cover the structure, organization and properties of peptide–synthetic hybrid block copolymers that have been obtained via the ring-opening polymerization of α-amino acid N-carboxyanhydrides. Details on the synthesis of these hybrid block copolymers are described elsewhere in this volume [17]. Peptide–synthetic hybrid block copolymers can also be prepared using solid phase peptide synthesis or using protein engineering techniques; the synthesis, structure and properties of these hybrid constructs are also discussed separately [18]. The remainder of this article will successively discuss five major classes of peptide–synthetic hybrid block copolymers, classified according to the chemical structure of the synthetic polymer block.

2
Polybutadiene-Based Block Copolymers

One of the first reports on the nanoscale solid-state structure of peptide–synthetic hybrid block copolymers was published by Gallot et al. in 1976 [19]. In this publication, the solid-state structure of a series of polybutadiene-b-poly(γ-benzyl-L-glutamate) (PB-PBLG) and polybutadiene-b-poly(N^5-hydroxylpropyl-L-glutamine) (PB-PHPG) diblock copolymers was investigated using a combination of techniques, including infrared and circular dichroism spectroscopy, X-ray scattering and electron microscopy. The block copolymers covered a broad composition range with peptide contents ranging from 19 to 75%. Interestingly, small angle X-ray scattering revealed a well-ordered lamellar superstructure characterized by up to four higher-order Bragg spacings for all of the investigated samples. The lamellar superstructure was confirmed by electron microscopy experiments, which were carried out on OsO_4-stained specimens. The intersheet spacings determined from the electron micrographs were in good agreement with the diffraction data. Wide angle X-ray scattering experiments indicated that the α-helical peptide blocks were assembled in a hexagonal array. For a number of block copolymer samples it was found that the calculated length of the peptide helix was larger than the thickness of the polypeptide layer. To accommodate this difference, it was proposed that the peptide helices were folded in the peptide layer. A schematic representation of the nanoscale structure of the PB-PBLG and PB-PHPG block copolymers as proposed by Gallot et al. is shown in Fig. 1. The lamellar structure consists of plane, parallel equidistant sheets. Each sheet is obtained by superposition of two layers; (i) the PB chains in a more or less random coil conformation and (ii) the α-helical polypeptide blocks in an hexagonal array of folded chains. The hexagonal-in-lamellar structure illustrated in Fig. 1 was also found for polybutadiene-b-poly(ε-benzyloxycarbonyl-L-lysine) (PB-PZLL) and polybutadiene-b-poly(L-lysine) (PB-PLL) block copolymers by the same authors [20–23]. In the case of the PB-PLL block copolymers no periodic arrangement of the poly(L-lysine)

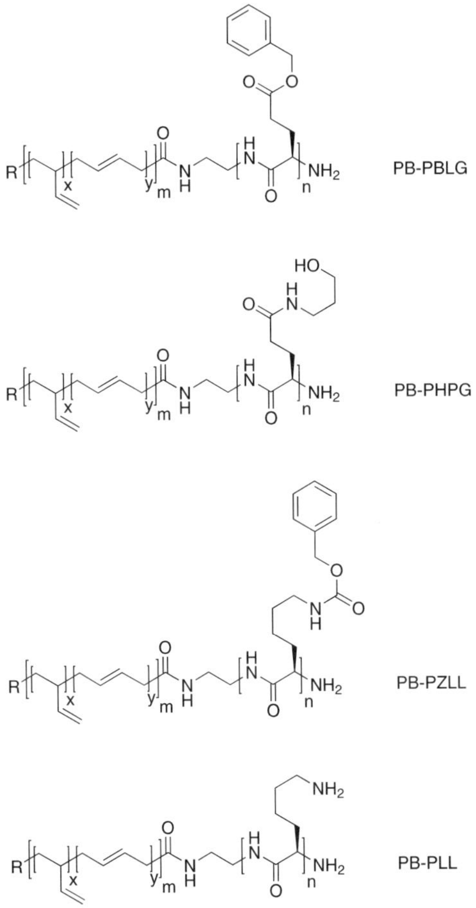

Scheme 1 Chemical structures of polybutadiene-b-poly(γ-benzyl-L-glutamate) (PB-PBLG), polybutadiene-b-poly(N^5-hydroxypropyl-L-glutamine) (PB-PHPG), polybutadiene-b-poly(ε-benzyloxycarbonyl-L-lysine) (PB-PZLL) and polybutadiene-b-poly(L-lysine) (PB-PLL)

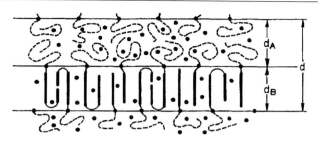

Fig. 1 Schematic representation of the lamellar microphase-separated structure of PB-PBLG block copolymers as proposed by Gallot et al. [19]. d_A = thickness of the PB layer; d_B = thickness of the PBLG layer. (Reproduced with permission from [19]. Copyright 1976 Wiley)

chains in the peptide layer was found. This is because of the fact that the polypeptide segments in these block copolymers were not exclusively α-helical but were composed roughly of 50% random coil, 35% α-helix and 15% β-strand domains [20].

Whereas Gallot et al. mainly studied the solid-state organization of PB-based diblock copolymers, the group around Nakajima has concentrated on ABA-type peptide–synthetic hybrid triblock copolymers containing polybutadiene as the B component. In a first series of publications, the structure and properties of poly(γ-benzyl-L-glutamate)-b-polybutadiene-b-poly(γ-benzyl-L-glutamate) triblock copolymers containing 7.5–32.5 mol % (= 3.0–14.3 vol %) polybutadiene were investigated [24–26]. Infrared spectroscopy experiments on films of the triblock copolymers indicated that the PBLG blocks were predominantly α-helical with helix contents close to that of a PBLG homopolymer. Wide angle X-ray scattering (WAXS) yielded an interhelix spacing of 12.5 Å, similar to PBLG. From the WAXS experiments, it was concluded that the PBLG blocks assembled into different structures, depending on the type of solvent that was used to cast the films. In benzene-cast films, the peptide helices were relatively poorly ordered. The WAXS patterns of these films were similar to the so-called form A morphology of PBLG [27]. In contrast, the PBLG segments in films cast from CHCl$_3$ were well-ordered and contained paracrystalline and mesomorphic regions. On the basis of transmission electron micrographs, a cylindrical microstructure was proposed for a triblock copolymer containing 8 vol % PB. Electron micrographs for other samples were not reported, but based on volume fraction considerations it was predicted that triblock copolymers containing 12 and 14 vol % PB would form either cylindrical or lamellar superstructures [25].

The microphase-separated structures formed by the PB-based γ-benzyl-L-glutamate containing triblock copolymers appeared to be sensitive to the conformation of the peptide blocks [28]. When instead of γ-benzyl-L-glutamate N-carboxyanhydride an equimolar mixture of γ-benzyl-L-glutamate N-carboxyanhydride and γ-benzyl-D-glutamate N-carboxyanhydride was used for the synthesis, the α-helical secondary structure of the peptide block was disrupted and the infrared spectra pointed towards the presence of a significant fraction of polypeptide blocks with a random coil conformation. The reduced helix content prevented a regular organization of the peptide segments (as evidenced from the WAXS patterns) and also influenced the microphase-separated structure. On the basis of transmission electron micrographs, a cylindrical or lamellar morphology was proposed for a triblock copolymer exclusively composed of γ-benzyl-L-glutamate. In contrast, a more spherical superstructure was suggested for a triblock copolymer of the same overall composition but with peptide blocks consisting of equal amounts of γ-benzyl-L-glutamate and γ-benzyl-D-glutamate.

Further support for the microphase-separated structure of the PBLG-PB-PBLG triblock copolymers was obtained from dynamic mechanical spectroscopy and water permeability experiments [26]. The temperature dependence of the dynamic modulus and the loss modulus could be well explained by assuming a microphase-separated structure. Furthermore, the hydraulic permeability of water through membranes prepared from the PBLG-PB-PBLG triblock copolymers was approximately three orders of magnitude larger compared to a pure PBLG membrane. The hydraulic water permeability was found to increase with increasing PB content in the block copolymers. The water permeability of the triblock copolymer membranes was explained in terms of their microphase-separated structure and the presence of an interfacial zone that separates the ordered domains formed by the α-helical PBLG chains from the unordered PB phase (Fig. 2). The interfacial zone consists of amino acid residues that are located close to the N-terminus of the peptide block and in the vicinity of the PB segment. The amino acid residues in the interfacial zone do not form regular secondary structures. Since the amide groups of the peptide chains in the interfacial zone are not involved in intramolecular hydrogen bonding, they are able to bind water molecules. Consequently, increasing the interfacial zone, for example by increasing the PB content, leads to an increase in the water permeability.

The bulk and surface structure of solvent-cast films from a series of PBLG-PB-PBLG triblock copolymers with much higher PB contents (50–80 mol %) than the samples discussed above have been discussed by Gallot et al. [29]. The organization of these PBLG-PB-PBLG triblock copolymers was compared with that of three other triblock copolymers with approximately the same PB content (~ 50 mol %) but which were composed of poly(N^{ε}-trifluoroacetyl-L-lysine), poly(N^5-hydroxyethyl-L-glutamine) or polysarcosine as the peptide block. At these high PB contents, for all of the investigated samples X-ray scattering experiments indicated a hexagonal-in-lamellar bulk morphology. X-ray photoelectron spectroscopy (XPS) measurements revealed that for the triblock copolymers with hydrophobic peptide blocks (poly(γ-benzyl-L-glutamate) or poly(N^{ε}-trifluoroacetyl-L-lysine)) the surface composition was identical to that of the bulk of the sample. In contrast, the surfaces of

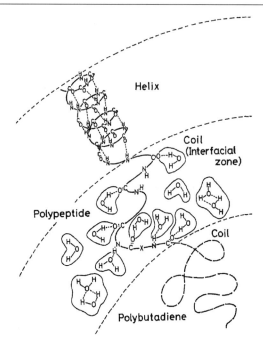

Fig. 2 Hydrogen-bonded water and water clusters in an interfacial zone formed by unordered peptide chains that separate the PB domains from the helical PBLG phase in films of PBLG-PB-PBLG triblock copolymers. (Reproduced with permission from [26]. Copyright 1979. The Society of Polymer Science, Japan)

films prepared from the triblock copolymers with the more hydrophilic peptide segments (poly(N^5-hydroxyethyl-L-glutamine) or polysarcosine) were PB enriched. Furthermore, the XPS data suggested that the lamellar superstructures formed by the triblock copolymers were perpendicular to the air–polymer interface.

In addition to the PBLG-PB-PBLG triblock copolymers discussed above, also the solid-state organization and properties of poly(ε-benzyloxycarbonyl-L-lysine)-b-polybutadiene-b-poly(ε-benzyloxycarbonyl-L-lysine) (PZLL-PB-PZLL) triblock copolymers have been investigated. Nakajima et al. have studied PZLL-PB-PZLL triblock copolymers composed of a central PB block with a molecular weight of 3600 and PB contents ranging from 12–52 mol % [30, 31]. WAXS patterns obtained from solution-cast triblock copolymer films were in agreement with the α-helical secondary structure of the peptide blocks. The bulk microphase-separated structure of the five different block copolymer samples could be successfully characterized by means of electron microscopy. For the samples with the largest PB volume fraction (56 and 65 vol %), a lamellar superstructure was found. For triblock copolymers with smaller PB volume fractions, the electron micrographs suggested cylindrical and spherical microphase separated structures.

To obtain information about the surface structure of the PZLL-PB-PZLL films, solvent-cast samples were investigated by means of XPS, replica electron microscopy and wettability experiments [31]. For the triblock copolymers with the smallest PB content (12 mol %), these experiments also confirmed a microphase-separated structure at the surface of the films. The PB content at the surface, however, was found to be about twice as large as the bulk PB content. To account for the enhanced PB content at the film surface, the PB domains were proposed to rise above the PZLL matrix in the form of convex lenses (Fig. 3). The presence of these convex PB domains at the film surface was confirmed by replica electron microscopy experiments. Unordered amino acid sequences form an interfacial zone at the film surface between the PB domains and the PZLL matrix. Since the amide bonds in the peptide sequences in the interfacial zone are not involved in intramolecular hydrogen bonding, they are able to bind water, which explains the low water contact angles that were measured on the PZLL-PB-PZLL films.

Other polybutadiene-based ABA-type triblock copolymers that have been investigated include poly(γ-methyl-L-glutamate)-b-polybutadiene-b-poly(γ-methyl-L-glutamate) (PMLG-PB-PMLG) and poly(γ-methyl-D,L-glutamate)-b-polybutadiene-b-poly(γ-methyl-D,L-glutamate) (PMDLG-PB-PMDLG) [28, 32]. Infrared spectroscopy experiments on solvent-cast films indicated that the incorporation of 50% of the D-isomer disrupts the α-helical secondary structure and induces a random coil conformation in significant portions of the peptide blocks. From the infrared spectra, it was estimated that the helix content of a poly(γ-methyl-D,L-glutamate) homopolypeptide was about 60% of that of the corresponding poly(γ-methyl-L-glutamate) homopolymer. The WAXS reflections recorded from the D,L triblock copolymers were much broader than those of the corresponding L-analogs. This also points towards a lower helix content and a less regular arrangement of the peptide blocks in the D,L triblock copolymer samples and is in agreement with the infrared spectroscopy experiments. Transmission electron micrographs of

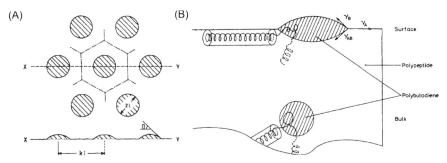

Fig. 3 Model of the surface topography of a PZLL-PB-PZLL film; **A** top and side view; **B** cross sectional representation. The shaded areas represent PB domains. (Reproduced with permission from [31]. Copyright 1982. The Society of Polymer Science, Japan)

OsO$_4$-stained samples provided evidence for the microphase-separated solid-state structure. Interestingly, different morphologies were observed when comparing the images of PMLG-PB-PMLG and PMDLG-PB-PMDLG samples with the same PB content ($\approx$ 30 mol %). For the PMLG-PB-PMLG triblock copolymer a cylindrical morphology and for the PMDLG-PB-PMDLG sample a spherical structure was proposed [28]. The difference in morphology was ascribed to the less regular secondary structure of the peptide block in the case of the D,L triblock copolymer, which prevents a highly ordered organization of the peptide domains and facilitates the formation of spherical PB domains. ATR-infrared spectroscopy experiments indicated that the PB content at the surface of the PMDLG-PB-PMDLG films was higher than the bulk concentration. On replica electron micrographs, the excess surface concentration was reflected in the form of convex PB domains, which were dispersed on top of a PMDLG matrix. The ATR-infrared spectra further showed that adsorption of bovine serum albumine (BSA) and bovine fibrinogen (BF) did not lead to denaturation. From these observations, the authors concluded that the surfaces of the PMDLG-PB-PMDLG membranes only weakly or reversibly interact with these plasma proteins and it was predicted that this may also lead to a good overall biocompatibility.

The solid-state structure and properties of poly(γ-ethyl-L-glutamate)-b-polybutadiene-b-poly(γ-ethyl-L-glutamate) (PELG-PB-PELG) triblock copolymers containing 31.5–94.5 mol % (= 17–88 vol %) PELG have been studied using the same techniques as described above for the other triblock copolymers [33, 34]. Infrared spectroscopy and WAXS experiments provided information on the peptide secondary structure and organization in solvent-cast films. The secondary structure of the PELG blocks was found to be predominantly α-helical and the helix content in the triblock copolymers decreased from 95 to 60% upon decreasing the peptide content from 95 to 61%. Interestingly, the WAXS data suggested that the PELG helices were packed in a pseudohexagonal, i.e. monoclinic, arrangement instead of the hexagonal structure observed for most of the other investigated peptide–synthetic hybrid block copolymers. Transmission electron microscopy experiments on OsO$_4$-stained films indicated a microphase-separated structure. On the basis of the electron micrographs, a spherical microphase-separated structure was proposed for the PELG-PB-PELG triblock copolymer containing 17 vol % PB, while cylindrical and lamellar morphologies were suggested for triblock copolymers containing 28 and 44 vol %, respectively, 68 and 88 vol % PB. The biocompatibility of the PELG-PB-PELG triblock copolymers was assessed by coating the samples onto a polyester mesh fiber cloth, which was subsequently subcutaneously implanted in mongrel dogs for 4 weeks. It was found that the foreign body reaction and degradation of the PELG-PB-PELG samples were less pronounced compared to PMLG-PB-PMLG, PBLG-PB-PBLG and PZLL-PB-PZLL triblock copolymers.

3
Polyisoprene-Based Block Copolymers

The bulk nanoscale structure of a series of poly(γ-benzyl-L-glutamate)-b-polyisoprene-b-poly(γ-benzyl-L-glutamate) (PBLG-PI-PBLG) triblock copolymers containing 37.4–81.1 mol % PBLG was studied by means of infrared spectroscopy, WAXS, dynamic mechanical analysis and electron microscopy [35]. The infrared spectroscopy experiments indicated that the PBLG blocks have an α-helical secondary structure and showed that the helix content (relative to that of a PBLG homopolymer sample) decreased from 76 to 34% upon decreasing the PBLG mol fraction from 81.1 to 37.4%. From WAXS an interhelical spacing of 12.5 Å was determined, in agreement with an hexagonal arrangement of the helical peptide blocks. Proof for the microphase-separated bulk morphology of the PBLG-PI-PBLG triblock copolymers was obtained from dynamic mechanical analysis and electron microscopy. On the basis of the electron micrographs, a cylindrical morphology was proposed for PBLG-PI-PBLG triblock copolymers containing 74.6 and 81.1 mol % PBLG. Water permeability measurements also supported the microphase-separated bulk morphology [36]. The hydraulic water permeability of membranes prepared by solvent casting of the PBLG-PI-PBLG triblock copolymers was three orders of magnitude larger than that measured across pure PBLG films. This was explained in terms of the unordered peptide sequences that are located near the N-terminus of the peptide blocks as well as close to the PI segment. As mentioned earlier, the amino acids in this interfacial region do not participate in helix formation and their amide bonds can bind water molecules via hydrogen bonding, which facilitates water transport across the membrane. Further insight into the bulk morphology of the PBLG-PI-PBLG triblock copolymers was obtained from pulsed proton NMR experiments [37]. The NMR signals of the block copolymers were composed of three components with different spin-spin relaxation times (T_2). The three different T_2's were attributed to the microphase-separated structure, which consists of three regions (the ordered helical peptide domains, the unordered interfacial peptide region and the rubbery PI phase) with different molecular mobility. The spin-lattice relaxation times (T_1) that were obtained provided insight into the domain sizes, which were in good agreement with the results from electron microscopy.

The surface structure of chloroform-cast PBLG-PI-PBLG was studied by XPS and contact angle measurements [38]. It was found that the chemical composition of the microphase-separated films at the surface was different from that in the bulk. The PI content at the film surface was higher than that in the bulk. Water contact angle measurements indicated that the block copolymer films were wetted easier than the respective homopolymers. This was explained in terms of the interfacial regions of unordered amino acid sequences, which were proposed to be located at the film surface. Finally,

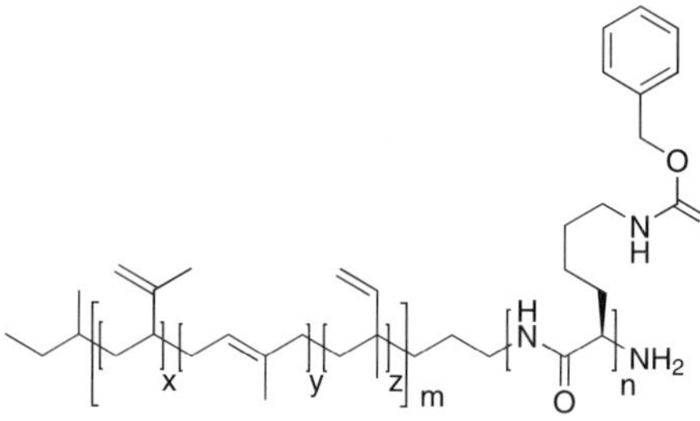

Scheme 2 Chemical structure of polyisoprene-b-poly(ε-benzyloxycarbonyl-L-lysine) (PI-PZLL)

in vitro blood and in vivo tissue compatibility experiments showed that the PBLG-PI-PBLG triblock copolymers had favourable biocompatibility properties compared to PI and PBLG homopolymer.

By treating a chloroform-cast PBLG-PI-PBLG film with a mixture of 3-amino-1-propanol and 1,8-octamethylenediamine, hydrophilic, crosslinked poly(N-hydroxypropyl-L-glutamine)-b-polyisoprene-b-poly(N-hydroxypropyl-L-glutamine) membranes were obtained [39]. The swelling ratio of these membranes in pseudoextracellular fluid (PECF) was found to decrease with increasing PI content and increasing crosslink density. Tensile tests in PECF revealed that the triblock copolymer membranes had a larger Young's modulus, increased tensile strength and elongation at break compared to membranes prepared from PBLG homopolymer. Enzymatic degradation experiments using papain showed that the triblock copolymer films were more resistant towards degradation than the corresponding homopolypeptide membranes. The half-times for sample degradation increased with decreasing peptide content, which was in agreement with the swelling behaviour of the membranes.

More recently, the solid-state nanoscale structure of polyisoprene-b-poly(ε-benzyloxycarbonyl-L-lysine) (PI-PZLL) diblock copolymers was reported [40]. Diblock copolymers composed of a PI block with a number-average degree of polymerization of 49 and a PZLL block containing 61–178 amino acid residues were investigated with dynamic mechanical analysis and X-ray scattering. For the PI_{49}-$PZLL_{35}$, PI_{49}-$PZLL_{61}$ and PI_{49}-$PZLL_{92}$, the X-ray scattering data were in agreement with an hexagonal-in-lamellar morphology. Interestingly, for PI_{49}-$PZLL_{92}$ the lamellar spacing was found to decrease when the samples were prepared from dioxane instead of THF/DMF and suggested folding of the peptide helices. For PI_{49}-$PZLL_{123}$ and PI_{49}-$PZLL_{178}$ a hexagonal-in-hexagonal structure was found. This morphology is illustrated in Fig. 4. This structure is unprecedented for polydiene-based peptide hybrid block copolymers, but has also been found for low molecular weight polystyrene-b-poly(γ-benzyl-L-glutamate) block copolymers (vide infra).

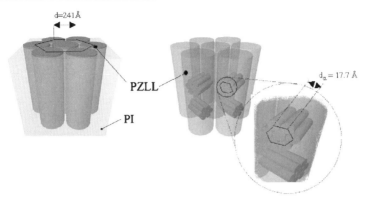

Fig. 4 Schematic representation of the hexagonal-in-hexagonal structure found in dioxane cast samples of PI$_{49}$-PZLL$_{178}$ [40]. (Reproduced by permission of The Royal Society of Chemistry)

4
Polystyrene-Based Block Copolymers

In an early study, Gallot et al. reported on the bulk nanoscale structure of polystyrene-b-poly(ε-benzyloxycarbonyl-L-lysine) (PS-PZLL) diblock copolymers, which were based on a polystyrene block with a number-average molecular weight of 37 000 g/mol and had peptide contents ranging from 18–80 mol % [21]. X-ray scattering patterns of dry samples that had been evaporated from dioxane showed two sets of signals. At very low angles, Bragg spacings characteristic of a layered superstructure were found. At somewhat larger angles, a second set of reflections pointing towards an hexagonal arrangement of peptide helices was found. On the basis of the X-ray data, a lamellar superstructure consisting of plane, parallel equidistant sheets was proposed. Each of the sheets results from the superposition of two layers; an amorphous PS layer and a peptide layer consisting of hexagonally packed peptide α-helices. For several samples, the calculated length of the peptide helix was larger than the peptide layer thickness as determined from the X-ray data. In these cases, it was proposed that the helical PZLL chains were folded in the peptide layer. Thus, the bulk nanoscale structure of the PS-PZLL diblock copolymers can be described in terms of the same hexagonal-in-lamellar model as was also proposed for the polybutadiene-based block copolymer described earlier (Fig. 1). Removal of the side-chain protective groups of the peptide segment resulted in polystyrene-b-poly(L-lysine) diblock copolymers [20]. The PS-PLL diblock copolymers were not water-soluble, but formed mesomorphic gels at water contents < 50%. The X-ray scattering patterns indicated a lamellar superstructure, both in the gel state and the dry samples. In contrast to the side-chain protected block copolymers, no evidence for a periodic arrangement of the peptide chains was found. This is not too surprising considering that IR spectra indicated that roughly 50% of the peptide blocks have a random coil conformation, 35% an α-helical secondary structure and 15% a β-strand conformation.

Along the same lines Douy and Gallot also studied the bulk nanoscale organization of polystyrene-b-poly(γ-benzyl-L-glutamate) (PS-PBLG) diblock copolymers [23]. For PS-PBLG block copolymers composed of a polystyrene block with a number-average molecular weight of 25 000 g/mol and containing 31–94 mol % peptide, the same hexagonal-in-lamellar morphology as described above for the PS-PZLL diblock copolymers was found. The biocompatibility of PS-PBLG block copolymers has been discussed in two publications [41, 42]. Mori et al. studied diblock copolymers composed of a polystyrene block with a number-average degree of polymerization of 87 and PBLG segments with number-average degrees of polymerization of 23, 52 or 83. Thrombus formation was assessed by exposing films of the diblock copolymers and the corresponding homopolypeptides to fresh canine blood. It was found that thrombus formation on the diblock copolymer films was reduced compared to the corresponding homopolymers. For the block copolymers, thrombus formation decreased with decreasing PBLG block length. Also, adsorption of plasma proteins such as BSA, bovine γ-globulin and bovine plasma fibrinogen was reduced on the block copolymers compared to polystyrene homopolymer.

Tanaka et al. studied ABA-type triblock copolymers composed of a central polystyrene block (B) flanked by two polypeptide segments [43]. Poly(γ-benzyl-L-glutamate) (PBLG), poly(ε-benzyloxycarbonyl-L-lysine) (PZLL) and polysarcosine were used as the polypeptide segment. Transmission electron micrographs of a CHCl$_3$-cast film of PBLG$_{25}$-PS$_{165}$-PBLG$_{25}$ that was stained with phosphotungstic acid revealed a lamellar phase-separated structure. In contrast, no microphase separation was observed in a CHCl$_3$-cast film of polysarcosine$_{73}$-PS$_{421}$-polysarcosine$_{73}$. The authors proposed that the different block copolymer morphologies could be related to the different secondary structure of the peptide block; while the PBLG segments are predominantly helical, the polysarcosine may not form any regular secondary structure.

Fibrinogen adsorption on CHCl$_3$-cast block copolymer films was studied with ATR FTIR spectroscopy and compared with that on the corresponding homopolymer films [43]. It was found that fibrinogen adsorption on PS and polysarcosine homopolymer films and on polysarcosine-PS-polysarcosine triblock copolymer films led to denaturation of the protein. In contrast, protein adsorption on the microphase-separated PBLG-PS-PBLG surfaces was reported to stabilize the protein's secondary structure. Blood clotting tests suggested that thrombus formation was retarded compared to the respective homopolymers.

Samyn et al. extended the investigations of ABA triblock copolymers and studied the solid-state organization of three different poly(γ-benzyl-L-glutamate)-b-polystyrene-b-poly(γ-benzyl-L-glutamate) (PBLG-PS-PBLG)

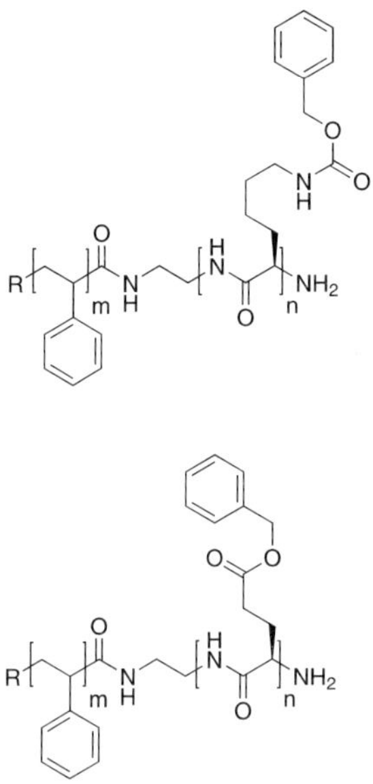

Scheme 3 Chemical structures of polystyrene-b-poly(ε-benzyloxycarbonyl-L-lysine) (PS-PZLL) and polystyrene-b-poly(γ-benzyl-L-glutamate) (PS-PBLG)

triblock copolymers containing 34, 55 and 92 wt % PBLG [44]. TEM micrographs of ultramicrotomed and RuO_4-stained specimens indicated a lamellar morphology for the 34 and 55 wt % triblock copolymers. For the triblock copolymer containing 92 wt % PBLG no lamellar structure was observed. Small angle X-ray scattering experiments provided additional evidence for the lamellar superstructure of the 34 and 55 wt % PBLG triblock copolymers. Wide angle X-ray scattering patterns yielded d-spacings reflecting the intermolecular distance between neighbouring peptide α-helices. Ion permeability measurements on dioxane-cast films indicated that the bulk morphology influenced the membrane properties [45]. The membranes prepared from the lamellae forming 34 and 55 wt % PBLG containing triblock copolymers showed cation selectivity. In contrast, the membrane prepared from the triblock copolymer containing 92 wt % PBLG did not show such a selectivity. It was proposed that uptake of cations into the triblock copolymer membranes was facilitated by the interactions between the cations and the ester functions in the block copolymers. The difference in selectivity between the lamellar 34 and 55 wt % PBLG triblock copolymers and the non-lamellar 92 wt % PBLG triblock copolymer was explained in terms of the interfacial zone that separates the PS and PBLG domains in films generated from the former two triblock copolymers. This interfacial region, as discussed earlier, consists of unordered peptide sequences with amide bonds that are available to interact with cations and can further facilitate their uptake. The membrane prepared from the 92 wt % PBLG triblock copolymer was assumed to have a much lower interfacial zone content.

The characterization of the solid-state nanoscale organization of polystyrene-based peptide–synthetic hybrid block copolymers was refined in a series of publications by Schlaad et al. In a first report, three polystyrene-b-poly(ε-benzyloxycarbonyl-L-lysine) (PS-PZLL) diblock copolymers with peptide volume fractions of 0.48, 0.74 and 0.82 were investigated [46]. Small-angle X-ray scattering patterns recorded from DMF-cast films supported the findings published earlier by Gallot et al. [21]. At lower scattering vectors two peaks in a ratio of 1 : 2 were observed, indicative for a lamellar superstructure. At larger scattering vectors, a set of three signals in the ratio of $1 : \sqrt{3} : 2$ was recorded, which is due to the hexagonal arrangement of the peptide helices in the peptide layer. Consequently, these studies confirmed the hexagonal-in-lamellar morphology proposed by Gallot et al. In their paper, Schlaad and coworkers, however, went a step further and analyzed their SAXS data using the interface distribution concept and the curvature-interface formalism. These evaluation techniques suggested that the bulk nanoscale structure of the PS-PZLL diblock copolymers does not consist of plain but rather undulated parallel lamellae (Fig. 5). The concept of the interface distribution function and the curvature-interface formalism were also applied to compare the solid-state structure of two virtually identical PS-based diblock copolymers; PS_{52}-$PZLL_{111}$ ($\phi_{peptide} = 0.82$) and PS_{52}-$PBLG_{104}$ ($\phi_{peptide} = 0.79$) [47]. Analysis of the SAXS data obtained on DMF-cast films indicated an hexagonal-in-undulated (or zigzag) lamellar morphology for both block copolymers. However, the X-ray data also revealed two striking differences between the samples. The first difference regards the thickness of the layers, which are a factor of three smaller for PS_{52}-$PBLG_{104}$ compared to PS_{52}-$PZLL_{111}$. Whereas the PZLL helices are fully stretched, the PBLG helices are folded twice in the layers. Since peptide folding increases the area per chain at the PS-PBLG interface, the thickness of the PS layers also has to decrease in order to cover the increased interfacial area. The second difference between the solid-state structures of PS_{52}-$PBLG_{104}$ and PS_{52}-$PZLL_{111}$ concerns the packing of the peptide helices. For the PZLL-based diblock copolymer it was estimated that ~ 180 peptide helices form an ordered domain. The level of ordering, however, was considerably lower for the peptide blocks of PS_{52}-$PBLG_{104}$ and only ~ 80 helices were estimated to form a single hexagonally ordered domain. The decreased order in the peptide layers of PS_{52}-$PBLG_{104}$ was suggested to be due to difficulties in aligning folded helices in a perfectly parallel hexagonal fashion. In addition, also the influence of the polydispersity of the polypeptide block on the solid-state morphology of PS-PZLL diblock copolymers has been studied [48]. To this end, a series of five PS-PZLL diblock copolymers was prepared from an identical ω-amino-terminated polystyrene macroinitiator (number-average degree of polymerization = 52, $M_w/M_n = 1.03$). The peptide content in these diblock copolymers varied between 0.43 and 0.68 and their polydispersity ranged from ~ 1.03–1.64. Evaluation of the SAXS data with the interface distribution function and the curvature-interface formalism confirmed, as expected, the hexagonal-in-undulated (or zigzag) lamellar solid-state morphology. Fractionation of the peptide helices according to their length leads (locally) to the formation of an almost plane, parallel lamellar interface, which is disrupted by kinks (undulations). The curvature at the PS-PZLL interface, however, was found to be strongly dependent on the chain length distribution of the peptide block. Block copolymers with the smallest molecular weight distribution produced lamellar structures with the least curvature. Increasing the chain length distribution of the peptide block (block copolymers with $M_w/M_n =\sim 1.25$) leads to larger fluctuations in the thickness of the PZLL layers, which increases the number of kinks and the curvature at the lamellar interface. At even larger polydispersities (~ 1.64), however, the number of kinks decreases again. With increasing polydispersity of the peptide block, the thickness fluctuations become larger and larger, as does the interfacial area. At a certain point, at sufficiently high polydispersity, the system tries to compensate for the increased interfacial tension and minimizes the number of kinks. The effect of the chain length distribution of the peptide block on the structure of the lamellar morphology of the PS-PZLL diblock copolymers is schematically illustrated in Fig. 6.

The examples discussed so far all involved relatively high molecular weight diblock copolymers. In this case, the molecular weight of the polypeptide block is usually sufficiently high so that it forms a stable α-helix and the common hexagonal-in-lamellar morphology is found. The situation changes, however, when the molecular weight of the block copolymers is significantly decreased. The influence of molecular weight on the solid-state organization of polystyrene-based peptide–synthetic hybrid block copolymers has been studied for a series of low molecular weight polystyrene-b-poly(γ-benzyl-L-glutamate) (PS-PBLG) and polystyrene-b-poly(ε-benzyloxycarbonyl-L-lysine) (PS-PZLL) diblock copolymers [49, 50]. These diblock copolymers

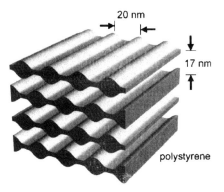

20 nm

17 nm

polystyrene

Fig. 5 Schematic representation of the (hexagonal-in-)undulated lamellar bulk nanoscale structure found for PS-PZLL diblock copolymers. (Reprinted from [46]. Copyright 2002, with permission from Elsevier)

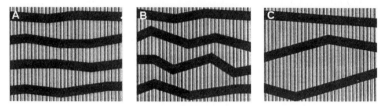

Fig. 6 Illustration of the undulated lamellar superstructure formed by PS-PZLL diblock copolymers with low (**A**), moderately (**B**) and highly polydisperse (**C**) polypeptide blocks. (Reprinted with permission from [48]. Copyright 2004. American Chemical Society)

consisted of a short polystyrene block with a number-average degree of polymerization of ~ 10, a polypeptide block containing ~ 10 to 80 amino acid repeat units and were characterized by means of variable temperature infrared spectroscopy and X-ray scattering. These experiments allowed the construction of "phase diagrams", which are shown in Fig. 7 for the PS-

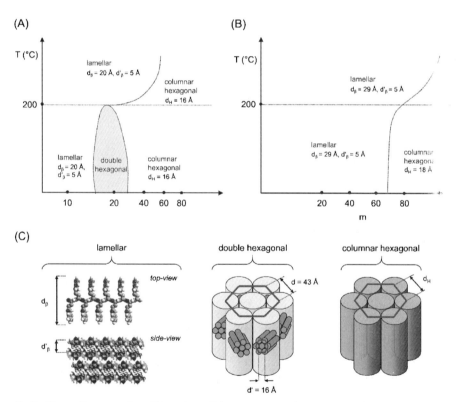

Fig. 7 Phase diagrams describing the solid-state nanoscale structure of **A** PS-PBLG and **B** PS-PZLL diblock copolymers; **C** Illustration of the lamellar, double hexagonal and hexagonal morphologies found for the low molecular weight hybrid block copolymers

PBLG and PS-PZLL diblock copolymers. The phase diagrams reveal a number of interesting features. At temperatures below $200\,^\circ$C and for sufficiently long polypeptide blocks a hexagonal arrangement of the diblock copolymers was found, analogous to the hexagonal-in-lamellar morphology of the high molecular weight analogues. Upon decreasing the length of the peptide block, however, several novel solid-state structures were discovered. For very short peptide block lengths (PS_{10}-$PBLG_{10}$, PS_{10}-$PZLL_{20}$, PS_{10}-$PZLL_{40}$ and PS_{10}-$PZLL_{60}$) a lamellar supramolecular structure was found. This is due to the fact that for such short peptide block lengths a substantial fraction of the peptide blocks adopts a β-strand secondary structure. Self-assembly of these diblock copolymers in a β-sheet type fashion results in the lamellar structures observed by SAXS. For PS_{10}-$PBLG_{20}$ a peculiar and until then unprecedented structure was found. This structure, which consists of hexagonally packed diblock copolymer molecules that are organized in a hexagonal superlattice, has been referred to as the double hexagonal or hexagonal-in-hexagonal morphology. Apart from several unconventional solid-state nanoscale structures, another factor that distinguishes the phase diagrams in Fig. 7 from those of most conventional, conformationally isotropic block copolymers is the influence of temperature. For a number of diblock copolymers, increasing the temperature above $200\,^\circ$C results in a change from an hexagonal-in-hexagonal (PS_{10}-$PBLG_{20}$) or hexagonal (PS_{10}-$PBLG_{40}$, PS_{10}-$PZLL_{80}$) to a lamellar morphology. FTIR spectroscopy experiments suggested that these morphological transitions are induced by an increase in the fraction of peptide blocks that has a β-strand conformation.

5
Poly(ethylene glycol)-Based Block Copolymers

Peptide–poly(ethylene glycol) (PEG) block copolymers are of particular interest, both from a structural and a functional point of view. Poly(ethylene glycol) is also often referred to as poly(ethylene oxide) (PEO). Throughout this article, however, this polyether will be referred to as PEG. In contrast to the hybrid block copolymers discussed in the previous paragraphs, which were based on amorphous synthetic polymers, PEG is a semi-crystalline polymer. In addition to microphase separation and the tendency of the peptide blocks towards aggregation, crystallization of PEG introduces an additional factor that can influence the structure formation of these hybrid block copolymers. Furthermore, PEG is an FDA approved biocompatible polymer, which makes peptide–PEG hybrid block copolymers potentially interesting materials for biomedical applications.

Inoue et al. have studied the adhesion behaviour of rat lymphocytes on solvent-cast films of poly(γ-benzyl-L-glutamate)-b-poly(ethylene glycol)-b-poly(γ-benzyl-L-glutamate) (PBLG-PEG-PBLG) triblock copolymers [51, 52].

The triblock copolymers were prepared from α,ω-bis-amino-functionalized poly(ethylene glycol) macroinitiators with molecular weights of 1000 and 4000 and had PEG contents varying from 11–33 wt %. Rat lymphocyte adhesivity was found to decrease with increasing PEG content. At the same PEG content, the adhesivity of the triblock copolymers based on the macroinitiator with a molecular weight of 4000 was lower than that of samples based on the macroinitiator with a molecular weight of 1000. In addition to overall lymphocyte adhesivity, the authors also studied the adhesion of specific subpopulations; B-cells and T-cells. All triblock copolymers showed a preference towards B-cells. However, whereas the B-cell/T-cell selectivity (A_B/A_T) of the PEG 4000-based triblock copolymers increased with increasing PEG content, A_B/A_T for the PEG 1000-based materials was independent of PEG content and remained constant at ~ 1.5. The B-cell/T-cell selectivity suggests that the triblock copolymers may be interesting candidates for the chromatographic separation of lymphocyte subpopulations without requiring the use of for example adjuvant proteins. The influence of PEG content on the cell adhesion behaviour was attributed to the microphase-separated structure of the triblock copolymer films and the changes in the relative sizes of the hydrophilic and hydrophobic domains that occur as the composition of the block copolymer is changed. The solvent that was used to cast the films was found to have a strong influence on the cell adhesion behaviour. For a PEG 4000-based triblock copolymer with a PEG content of 26 wt %, the adhesivity and A_B/A_T were $28 \pm 2\%$ / 2.47 ± 0.28, $17 \pm 1\%$ / 2.99 ± 0.64 and $70 \pm 3\%$ / 1.23 ± 0.03 when THF, benzene, respectively DMF were used. In an attempt to explain these differences, circular dichroism spectra were recorded and water contact angles measured. These experiments, however, revealed that the observed differences in cell adhesion behaviour were neither due to differences in the conformation of the peptide blocks, nor could be attributed to differences in surface hydrophilicity. It was therefore proposed that the observed effects were caused by differences in the higher order surface structures, i.e. in terms of the microphase-separated morphology and/or PEG crystallinity.

Kugo et al. studied the solid-state conformation of the peptide segment of a series of poly(γ-benzyl-L-glutamate)-b-poly(ethylene glycol)-b-poly(γ-benzyl-L-glutamate) (PBLG-PEG-PBLG) triblock copolymers containing a PEG segment with a molecular weight of 4000 and 36–86 mol % PBLG [53]. Infrared spectroscopy experiments on $CHCl_3$-cast films revealed that the PBLG blocks, which had degrees of polymerization of 25–276, had an α-helical secondary structure. The helix content of the triblock copolymer containing PBLG blocks with 276 repeat units was estimated to be similar to that of the PBLG homopolymer. Swelling the triblock copolymer films with water resulted in a decrease in helix content, as indicated by the CD spectra. This decrease in helicity was attributed to competition of water clusters to form hydrogen bonds with the peptide backbone. The effect was even more pronounced when pseudo-extracellular fluid was used instead of water.

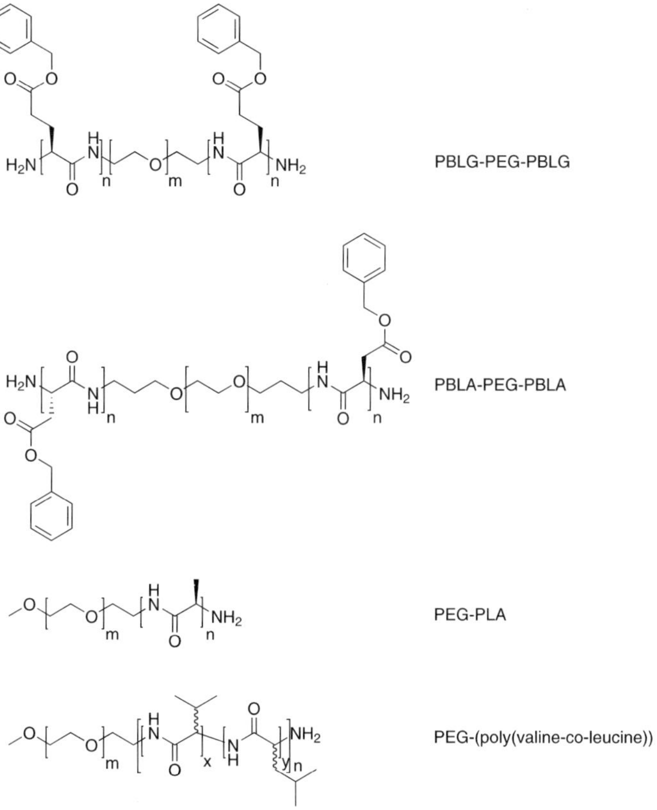

Scheme 4 Chemical structures of poly(γ-benzyl-L-glutamate)-b-poly(ethylene glycol)-b-poly(γ-benzyl-L-glutamate) (PBLG-PEG-PBLG), poly(γ-benzyl-L-asparate)-b-poly(ethylene glycol)-b-poly(γ-benzyl-L-asparate) (PBLA-PEG-PBLA), poly(ethylene glycol)-b-poly(L-alanine) (PEG-PLA) and poly(ethylene glycol)-b-poly(DL-valine-co-DL-leucine) (PEG-poly(valine-co-leucine))

A first detailed study of the solid-state nanoscale structure of peptide–PEG hybrid block copolymers was published by Cho et al. [54]. These authors investigated thin, chloroform-cast films of PBLG-PEG-PBLG triblock copolymers, which were composed of a PEG block with a number-average molecular weight of 2000 and contained 25–76 mol % PBLG. TEM micrographs of RuO$_4$-stained specimens revealed a lamellar microphase-separated morphology for triblock copolymers containing 25–64 mol % PBLG. The microphase-separated structure was proposed to consist of chain folded, crystalline PEG domains and helical PBLG domains. The helical secondary structure of the PBLG segments was confirmed by IR spectroscopy. Wide-angle X-ray scattering patterns obtained from CHCl$_3$-cast films were consistent with the ordered, crystalline-like solid-state modification C of PBLG. In contrast, in benzene-cast films, the peptide blocks only formed poorly ordered arrays.

The sensitivity of the organization of the PBLG blocks towards the nature of the casting solvent is identical to the behaviour of the PBLG homopolymer.

In a separate study, the degradation behaviour of PBLG-PEG-PBLG triblock copolymers was investigated [55]. The sensitivity of the triblock copolymers towards enzymatic degradation was investigated by incubating films in solutions containing the enzyme protease and measuring weight loss as a function of time. The rate of degradation was found to increase with increasing PEG content in the triblock copolymers from 1.4 to 3.1 to 13.6 mol %. For the degree of swelling a similar dependence on PEG content was observed. Exposure of the triblock copolymer samples to a PBS solution without the enzyme did not result in measurable weight loss, indicating that hydrolytic degradation did not take place.

While the data reported by Cho et al. described the structure and organization of thin solvent-cast films of PBLG-PEG-PBLG triblock copolymers, Floudas and coworkers have extensively studied the bulk nanoscale organization of these materials [56]. To this end, a series of PBLG-PEG-PBLG triblock copolymers with PBLG volume fractions (f_{PBLG}) ranging from 0.07–0.89 was investigated using X-ray scattering, polarizing optical microscopy, DSC and FTIR spectroscopy. For triblock copolymers with $f_{PBLG} \leq 0.25$, PEG crystallization was observed, however, with significant undercooling. As an example, for the triblock copolymer with $f_{PBLG} = 0.17$ a melting point hysteresis of 52 K was measured. Triblock copolymers with $f_{PBLG} \geq 0.43$ did not show PEG crystallization. SAXS experiments, which were carried out at 373 K (i.e. above the melting point of PEG) also revealed a different behaviour for triblock copolymers with small and large PBLG volume fractions. For triblock copolymers with $f_{PBLG} \geq 0.43$ only a weakly phase-separated structure was found, whereas for samples with $f_{PBLG} \leq 0.25$ the SAXS data clearly indicated a microphase-separated structure. WAXS patterns showed that in the microphase-separated state the PEG phase was semi-crystalline and the peptide phase consisted of hexagonally ordered assemblies of PBLG α-helices that coexisted with β-sheet structures. For triblock copolymers with $f_{PBLG} \geq 0.43$, PEG is amorphous and interspersed with aggregates of α-helical PBLG segments and unordered peptide chains. These different bulk structures are schematically illustrated in Fig. 8. This figure illustrates how the competing interactions that promote the bulk self-assembly of the PBLG-PEG-PBLG triblock copolymers lead to the formation of hexagonally ordered structures that cover different length scales. At the smallest length scale, hydrogen bonding interactions stabilize peptide secondary structures (α-helices, β-strands) and PEG chain folding occurs. On the next higher level, peptide α-helices and β-strands form hexagonal assemblies, respectively β-sheet structures. Finally, the mutual incompatibility of the peptide and PEG block leads to microphase separation.

Additional insight into the solid-state nanoscale organization of the series of PBLG-PEG-PBLG triblock copolymers discussed in the previous paragraph was obtained by combining SAXS/WAXS with different microscopic

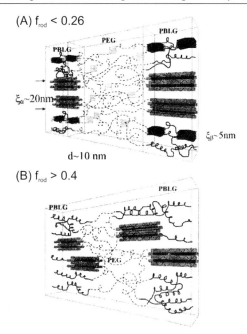

Fig. 8 Model description of the nanoscale solid-state structure of PBLG-PEG-PBLG triblock copolymers with PBLG volume fractions < 0.26 (**A**) and > 0.4 (**B**). (Reprinted with permission from [56]. Copyright 2003. American Chemical Society)

techniques (TEM, AFM) [57]. WAXS experiments on $CHCl_3$-cast samples revealed PEG crystallization for triblock copolymers with $f_{PBLG} \leq 0.26$. The observed reflections were in agreement with a monoclinic unit cell of 7_2 helical PEG. For triblock copolymers with $f_{PBLG} \geq 0.3$, no crystalline PEG reflections were observed in the WAXS patterns. The WAXS experiments also indicated that for PBLG blocks containing less than 18 repeat units, the β-strand conformation was predominant. In triblock copolymers with longer PBLG segments, the major fraction of the peptide block had an α-helical secondary structure. The peptide helices, however, were found to possess helical order only over short distances. Annealing the samples at 100 °C overnight promoted the formation of α-helical secondary structures. TEM experiments were carried out on thin $CHCl_3$-cast films that were stained with uranyl acetate and subsequently annealed at 100 °C overnight. For $PBLG_{16}$-PEG_{138}-$PBLG_{16}$ ($f_{PBLG} = 0.17$) and $PBLG_{17}$-PEG_{90}-$PBLG_{17}$ ($f_{PBLG} = 0.25$), the TEM micrographs suggested a morphology composed of PBLG cylinders embedded in a PEG matrix. The PBLG cylinders were proposed to consist of bilayers of β-strand peptides. TEM micrographs of triblock copolymers with larger PBLG volume fractions revealed enhanced ordering. For $PBLG_5$-PEG_{21}-$PBLG_5$ ($f_{PBLG} = 0.295$) and $PBLG_9$-PEG_{21}-$PBLG_9$ ($f_{PBLG} = 0.43$) a lamellar morphology was observed. The measured thickness of the PBLG layers was consistent with a bilayer of β-sheets and a layer of amorphous PEG. A "broken lamellar" morphology was observed in the TEM micrographs of PBLG$_{58}$-PEG$_{90}$-PBLG$_{58}$ (f_{PBLG} = 0.58). Annealing converted this metastable structure into a non-uniform microphase-separated pattern, which was proposed to consist of "puck-like" PEG domains in a PBLG matrix. For PBLG$_{105}$-PEG$_{90}$-PBLG$_{105}$ (f_{PBLG} = 0.67), a lamellar morphology was found in the as-cast film, which was transformed into a broken lamellar structure upon annealing. AFM experiments supported the TEM observations. On the basis of the results, a morphology map was constructed (Fig. 9). Figure 9 shows disordered phases at very low and very high peptide volume fractions, as expected based on the usual upper critical ordering transition temperature behaviour of diblock copolymers. With increasing f_{PBLG}, the following sequence of morphologies was observed: rods, lamellae, broken lamellae/pucks.

More complex, four-arm star block copolymer architectures were obtained by ring-opening polymerization of γ-benzyl-L-glutamate N-carboxyanhydride using the tetrafunctional bis(poly(ethylene glycol)bis(amine)) (M = 20 000) as the macroinitiator [58]. The resulting star block copolymers contained 14.9, 33.3 and 40.3 mol % PBLG. DSC experiments yielded a melting temperature that was slightly lower than that of pure PEG. The crystallinity of the PEG segments, however, was found to strongly decrease with decreasing PEG content. This was consistent with the results from WAXS experiments, which showed a decrease in the intensity of the crystalline PEG reflections with decreasing PEG content. For the block copolymer containing 40.3 mol % PBLG, the crystalline PEG reflections were completely absent. Infrared spectra from CHCl$_3$-cast films indicated that the PBLG segments in the block copolymers have an α-helical secondary structure. The WAXS patterns revealed a reflection corresponding to a spacing of 12.5 Å, consistent with the

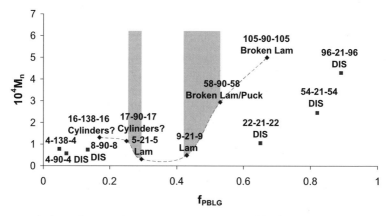

Fig. 9 Morphology diagram for PBLG$_n$-PEG$_m$-PBLG$_n$ triblock copolymers compiled on the basis of TEM, AFM and SAXS data. Subscripts m and n refer to the number-average degree of polymerization of the PEG and PBLG blocks, respectively

expected interhelical distance. TEM micrographs of RuO_4-stained specimens revealed a phase-separated structure consisting of globular PBLG domains surrounded by PEG for the star block copolymer containing 40.3 mol % PBLG. In the TEM micrograph of the 33 mol % PBLG sample, these globular structures were found to coexist with rod-like aggregates composed of a PBLG core covered with a PEG sheath. For the star block copolymers with the smallest PBLG content, rod-like aggregates were the predominant morphology. The rate of albumin release from disks prepared from tetra-arm PEG-PBLG block copolymers containing 19.1, 51.8 and 75.3 mol % PEG was found to increase with increasing PEG content [59]. These observations were explained by the increased degree of swelling of the disks with increasing PEG content.

In addition to PBLG-PEG-PBLG triblock copolymers, also poly(ε-benzyl-oxycarbonyl-L-lysine)-b-poly(ethylene glycol)-b-poly(ε-benzyloxycarbonyl-L-lysine) (PZLL-PEG-PZLL) triblock copolymers have been studied. Cho and colleagues reported on the solid-state structure of a series of PZLL-PEG-PZLL triblock copolymers composed of a PEG block with a number-average molecular weight of 2000 and PZLL contents of 25.2, 49.9 and 83.0 mol % (=68, 86 and 98 vol %) [60]. Infrared spectra of $CHCl_3$-cast films were in agreement with a helical secondary structure of the peptide blocks. DSC experiments provided a first hint for the existence of a microphase-separated structure and revealed two glass transition temperatures (T_g) for all samples. The higher T_g was very close to that of the PZLL homopolymer and the lower T_g approximately 20 °C higher than that of PEG homopolymer. A PEG melting transition was not observed. These results were interpreted in terms of a microphase-separated structure with hard, crystalline PZLL domains and soft, amorphous PEG segments. The presence of a microphase-separated structure was confirmed by TEM micrographs of RuO_4-stained thin films of the 25 and 50 mol % PZLL containing triblock copolymers.

Akashi et al. reported on the solid-state nanoscale structure of ABA-type triblock copolymers composed of a central PEG block flanked by two poly(β-benzyl-L-aspartate) blocks (PBLA-PEG-PBLA) [61]. The molecular weight of the central PEG block was 11 000 or 20 000 and the degrees of polymerization of the peptide blocks ranged from 12 to 32 repeat units. WAXS and polarizing optical microscopy studies on CH_2Cl_2-cast films showed PEG crystallization in all samples. The intensity of the crystalline PEG reflection peak, however, was found to decrease with increasing PBLA block length. The observation of PEG crystallization was interpreted as a first indication for microphase separation. In addition to the PEG signal, the WAXS patterns also contained reflections at $2\Theta = 5.9°$ (= 15 Å), which was assigned to a hexagonally packed array of PBLA helices. The helical secondary structure of the PBLA was confirmed by FTIR spectroscopy. In the SAXS patterns of $PBLA_{25}$-PEG_{250}-$PBLA_{25}$, $PBLA_{25}$-PEG_{454}-$PBLA_{25}$ and $PBLA_{32}$-PEG_{454}-$PBLA_{32}$ broad and weak diffraction peaks were observed, indicating the formation of phase-separated structures. The diffraction peaks measured for $PBLA_{25}$-PEG_{250}-

PBLA$_{25}$, PBLA$_{25}$-PEG$_{454}$-PBLA$_{25}$ and PBLA$_{32}$-PEG$_{454}$-PBLA$_{32}$ corresponded to periodicities of 150–70, 290–130 and 220–110 Å, respectively. Thermal analysis of the triblock copolymers, however, revealed several interesting properties. DSC experiments showed that the melting temperature of the crystalline PEG domains decreased linearly with increasing PBLA content, reflecting the strong influence of the peptide segments on PEG crystallization. More interestingly, the authors found that heating the as-cast films to > 333 K and cooling down to 303 K, converted a certain fraction of the α-helical PBLA chains into β-strands and was accompanied by a decrease in PEG crystallinity. On a macroscopic level, this led to increased strength and elasticity of the films.

Ma et al. have investigated the solid-state structure and properties of three poly(ethylene glycol)-b-poly(L-alanine) (PEG-PLA) diblock copolymers that were prepared from a PEG macroinitiator with a number-average molecular weight of 2000 [62]. The diblock copolymers contained 39.8, 49.6 and 65.5 mol% PLA. FTIR spectra of the diblock copolymers were indicative for an α-helical secondary structure of the peptide block. DSC traces of the polymers, in particular for the sample containing 49.6 mol% PLA, showed features of both of the respective homopolymers, which led the authors to propose a microphase-separated bulk structure.

AB diblock and ABA triblock copolymers composed of PEG as the A block and random coil segments of poly(DL-valine-co-DL-leucine) as the B block(s) were investigated by Cho, Kim and Jung [63]. The block copolymers were prepared by ring-opening copolymerization of DL-valine N-carboxyanhydride and DL-leucine N-carboxyanhydride using α-methoxy,ω-amino PEG or α,ω-diamino-PEG as macroinitiators. Since DL-valine N-carboxyanhydride and DL-leucine N-carboxyanhydride were used as monomers, the peptide segments were expected to have a random coil conformation. DSC experiments revealed PEG crystallization and showed that the PEG melting temperature was decreased compared to that of PEG homopolymer. TEM micrographs suggested a lamellar microphase-separated structure for one of the triblock copolymer samples.

6
Other Polyether-Based Block Copolymers

Cho et al. have studied triblock copolymers composed of a middle block of poly(propylene glycol) (PPG) with a molecular weight of 2000 flanked by two poly(γ-benzyl-L-glutamate) (PBLG) segments [64]. Three triblock copolymer samples were investigated with PPG contents of 17.0, 26.0 and 60.0 mol%, respectively. According to infrared spectra that were recorded from CHCl$_3$-cast films, the PBLG blocks possessed an α-helical secondary structure. WAXS patterns revealed a 12.5 Å interhelical spacing and were in agreement with a solid-state modification C of PBLG. No further details on the solid-state nanoscale structure and the possibility of microphase separation were reported.

Three-arm star poly(propylene glycol)-b-poly(γ-methyl-L-glutamate) (PPG-PMLG) block copolymers have been prepared using a triamine-functionalized PPG derivative to initiate the ring-opening polymerization of γ-methyl-L-glutamate N-carboxyanhydride [65]. Three different samples were studied, which contained 33.0, 47.0 and 70.0 mol % PPG, respectively. Infrared spectra obtained from TFE-cast films indicated that the PMLG blocks assume an α-helical secondary structure. WAXS measurements showed a main reflection that corresponded to the expected interhelical spacing of PMLG helices. Platelet adhesion on glass beads coated with block copolymers containing 33 or 47 mol % PPG was reduced compared to beads modified with PMLG homopolymer or block copolymers with 70.0 mol % PPG. These differences were attributed to differences in surface composition and morphology, which, unfortunately, were not further discussed.

Hayashi, Kugo and Nakajima have reported poly(γ-methyl-L-glutamate)-b-poly(tetrahydrofuran)-b-poly(γ-methyl-L-glutamate) (PMLG-PTHF-PMLG) triblock copolymers that were prepared from an amino-functionalized poly(tetrahydrofuran) macroinitiator with a molecular weight of 9600 [66]. Three different triblock copolymers were studied with PMLG contents and degrees of polymerization of 86.8 mol %/460, 89.7 mol %/605 and 91.3 mol %/730, respectively. With respect to the solid-state structure and organization, only infrared spectra and WAXS data were discussed. Infrared

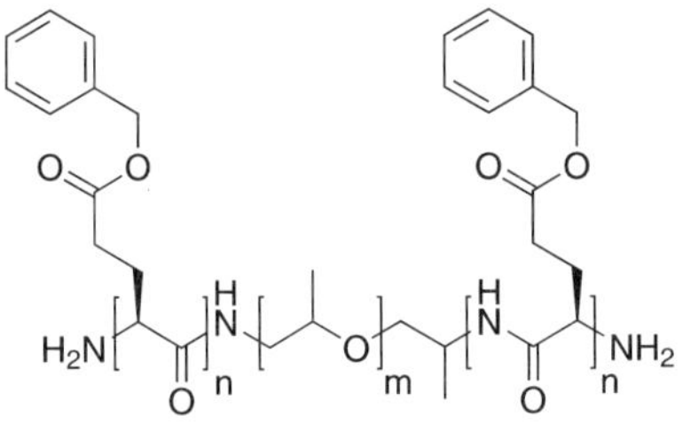

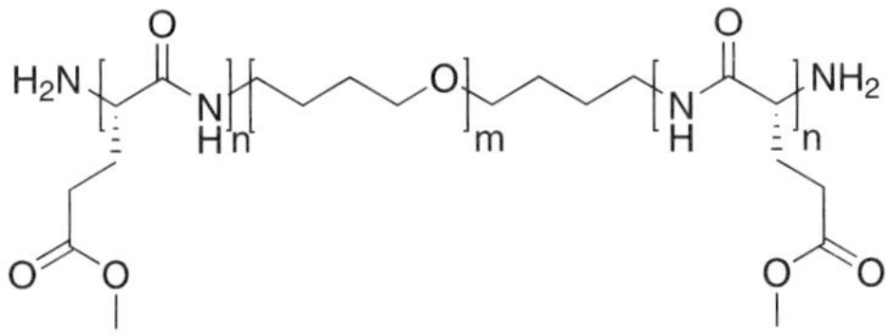

Scheme 5 Chemical structures of poly(γ-benzyl-L-glutamate)-b-poly(propylene glycol)-b-poly(γ-benzyl-L-glutamate) (PBLG-PPG-PBLG) and poly(γ-methyl-L-glutamate)-b-polytetrahydrofuran-b-poly(γ-methyl-L-glutamate) (PMLG-PTHF-PMLG)

spectra of solvent-cast films showed that the peptide blocks had an α-helical secondary structure. WAXS patterns supported these data and revealed a peak corresponding to a spacing of 10.34 Å, which is in good agreement with the expected interhelical spacing for PMLG.

7
Polydimethylsiloxane-Based Block Copolymers

Imanishi et al. have studied the structure, antithrombogenicity and oxygen permeability of ABA triblock copolymers composed of polydimethylsiloxane as the B block and poly(γ-benzyl-L-glutamate) (PBLG), poly(γ-benzyl-D,L-glutamate) (PBDLG), poly(ε-benzyloxycarbonyl-L-lysine) (PZLL) or polysarcosine as the A block [67]. Several series of triblock copolymers were prepared using bifunctional PDMS macroinitiators and targeting different peptide block lengths. TEM images of DMF-cast films provided evidence for a microphase-separated morphology for $PZLL_{49}$-$PDMS_{400}$-$PZLL_{49}$, $PZLL_{91}$-$PDMS_{256}$-$PZLL_{91}$ and $PZLL_{160}$-$PDMS_{256}$-$PZLL_{160}$. The images revealed a spherical morphology composed of PDMS islands in a PZLL matrix. The formation of these spherical domains was attributed to the solvent that was used for sample preparation. While DMF is a good solvent for PZLL, it is a poor solvent for PDMS. In a separate publication, the same authors also described non-spherical microphase-separated structures [68]. In CH_2Cl_2-cast films of a triblock copolymer with a very high PDMS content ($PBLG_{48}$-$PDMS_{508}$-$PBLG_{48}$, 83 mol % PDMS) more extended, rod-like PBLG aggregates in a matrix of PDMS were observed. The TEM experiments also provided insight into the effects of peptide secondary structure and the nature of the casting solvent on the thin film morphology [67]. Thin films of $PBLG_{42}$-$PDMS_{148}$-$PBLG_{42}$ prepared from DMF showed a spherical morphology. Changing the solvent from DMF (which is a good solvent for PBLG) to CH_2Cl_2 (which is a fairly non-selective solvent) resulted in coarsening of the microphase-separated structures. $PBDLG_{42}$-$PDMS_{148}$-$PBDLG_{42}$ films prepared from CH_2Cl_2 also showed a microphase-separated structure in which spherical PDMS domains were embedded in a PBDLG matrix. The dimensions of the spherical domains, however, were much smaller than those observed in the TEM images of $PBLG_{42}$-$PDMS_{148}$-$PBLG_{42}$. These different morphologies reflect the influence of the peptide secondary structure on the block copolymer self-assembly. While the peptide segments in $PBLG_{42}$-$PDMS_{148}$-$PBLG_{42}$ adopt an α-helical conformation, the PBLG blocks in $PBDLG_{42}$-$PDMS_{148}$-$PBDLG_{42}$ have a random coil conformation.

Blood clotting tests revealed that thrombus formation on the PBLG, PBDLG, PZLL and polysarcosine-based PDMS triblock copolymer films was reduced in comparison to glass surfaces and the respective homopolymers [67]. A systematic variation of block copolymer chain length and composition indicated that the best antithrombogenicity was obtained at polypeptide contents of 65–75 mol %, irrespective of the chemical composition of the block copolymer. It was also found that the antithrombogenicity of PBLG$_{56}$-PDMS$_{256}$-PBLG$_{56}$ was improved by hydrolysis of the ester groups to carboxylic acids and even further by converting these into the corresponding sodium carboxylates.

A detailed study on the adsorption of plasma proteins and platelet adhesion on films prepared from PBLG-PDMS-PBLG triblock copolymers was described by Imanishi et al. in a separate report [68]. The adsorption of plasma proteins was studied with FTIR spectroscopy. The adsorption of bovine γ globulin (BγG) was found to be independent of block copolymer composition. For bovine plasma fibrinogen (BPF) and BSA no clear correlation between protein adsorption and block copolymer composition was found. FTIR spectroscopy was not only used in an attempt to quantify protein adhesion, but also provided information on the conformation of the adsorbed proteins. It was found that BSA was not denatured on any of the investigated PBLG-PDMS-PBLG surfaces. BγG and BPF were completely denatured on surfaces prepared from block copolymers containing more than 70 mol % PDMS. Denaturation of BγG and BPF did not occur on films from triblock copolymers

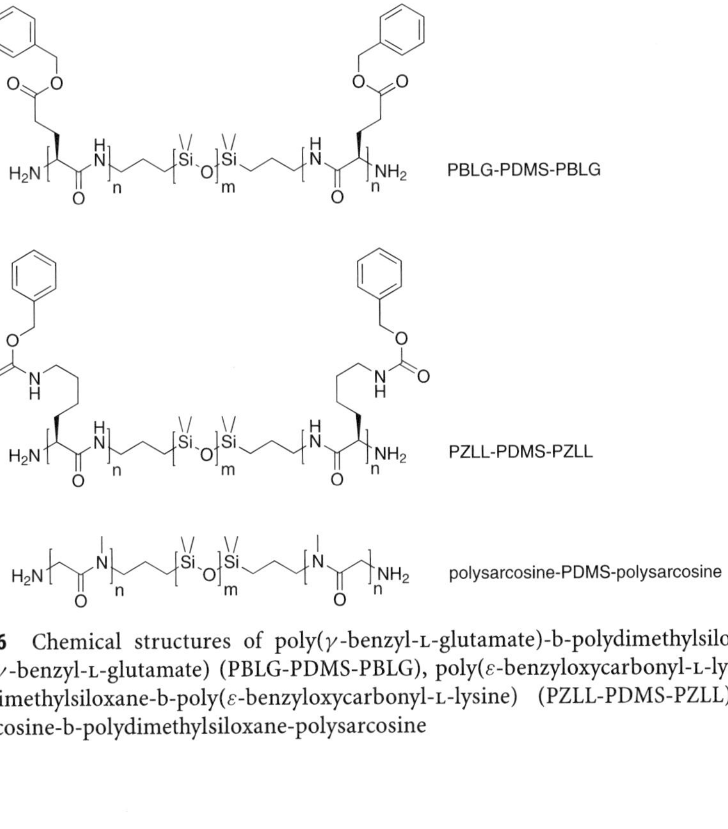

Scheme 6   Chemical structures of poly(γ-benzyl-L-glutamate)-b-polydimethylsiloxane-b-poly(γ-benzyl-L-glutamate) (PBLG-PDMS-PBLG), poly(ε-benzyloxycarbonyl-L-lysine)-b-polydimethylsiloxane-b-poly(ε-benzyloxycarbonyl-L-lysine) (PZLL-PDMS-PZLL) and polysarcosine-b-polydimethylsiloxane-polysarcosine

containing 50 mol % PDMS. For certain block copolymer compositions, it was found that alkaline hydrolysis of the benzyl ester groups allowed the prevention of denaturation. Systematic investigations revealed that denaturation did not occur on block copolymer films containing 40–70 mol % PDMS and water contact angles of 50–85°. The authors also observed that the number of adhered platelets and the amount of released serotonin increased on surfaces that induce protein denaturation. This finding was interpreted as an indication for the close relationship between protein adsorption–denaturation and platelet adhesion–activation.

Oxygen permeation measurements on CH_2Cl_2-cast peptide-PDMS-peptide triblock copolymer films showed that for samples with the same composition, the oxygen permeability decreased with increasing degree of polymerization of the PDMS (and consequently also of the polypeptide) block [67]. This was explained in terms of the size of the different domains in the microphase-separated structure. With decreasing length of the PDMS and polypeptide block, the size of the microphase-separated domains also decreases, leading to an enhanced interface area and promoting oxygen permeation. A similar observation was also made when the block copolymer films were prepared from DMF instead of CH_2Cl_2. DMF, being a poorer solvent for PDMS, leads to smaller microphase-separated structures and a higher oxygen permeability.

A detailed study of the gas permeation properties of CH_2Cl_2 and DMF-cast PBLG-PDMS-PBLG films with PDMS contents ranging from 46 to 83 mol % was reported separately [69]. The oxygen permeability of the triblock copolymer films in water was found to exponentially increase with increasing PDMS content. This is in agreement with a microphase-separated morphology of the membranes. Increasing the PDMS content increases the volume fraction of the gas permeable PDMS domains and promotes oxygen transport. The oxygen permeability across triblock copolymer membranes containing > 78 mol % PDMS was found to be enhanced in the dry state compared to the measurements that were performed in water. This effect was attributed to the hydrophobic surface properties of these PDMS-rich membranes, which was suggested to result in an increased interfacial resistance to oxygen transport. Similar results were reported by Kugo, Nishioka and Nishino, who studied oxygen and nitrogen transport across PBLG-PDMS-PBLG triblock copolymers containing 63–81 mol % PBLG [70].

8
Other Peptide–Synthetic Hybrid Block Copolymers

One of the first studies focussing on the solid-state properties of peptide-polyester synthetic hybrid block copolymers was reported by Jérôme et al. [71]. DSC experiments on a poly(ε-caprolactone)$_{50}$-b-poly(γ-benzyl-L-glutamate)$_{40}$ (PCL$_{50}$-PBLG$_{40}$) diblock copolymer revealed two endotherms.

The first endotherm was found at 60 °C and is due to the melting of the PCL. The second endotherm which was located at 110 °C, was, mistakenly, interpreted as the melting transition of PBLG. This transition, however, is not a melting transition, but instead reflects the conformational transition of the PBLG helix from a 7/2 to a 18/5 helical structure. Although no further structural investigations were carried out, the observation of two separate endotherms occurring at temperatures identical to the transitions found for the respective homopolymers is a first indication for the existence of a microphase-separated structure. Similar results were reported by Chen et al. who investigated the thermal properties of a series of PCL-PBLG block copolymers composed of PCL blocks containing 13–51 repeat units and peptide segments with 22–52 amino acid repeat units [72].

Caillol et al. have studied the solid-state structure and properties of a series of poly(L-lactic acid)-b-poly(γ-benzyl-L-glutamate) (PLLA-PBLG) diblock copolymers [73]. The PLLA block in these copolymers contained 10–40 repeat units and the peptide segments were composed of 20–100 repeat units. DSC thermograms of the block copolymers revealed three transitions corresponding to the glass transition temperature (T_g) of PLLA ($\sim 50\,°C$), the 7/2 to 18/5 helix transition of PBLG ($\sim 100\,°C$) and the melting temperature of PLLA ($\sim 160\,°C$), respectively. This observation already provided a first hint towards a microphase-separated bulk morphology. SAXS experiments, which were performed at 100 °C, indicated the existence of hexagonally ordered assemblies of α-helical PBLG chains. With decreasing PBLG content, the peaks corresponding to this hexagonal organization decreased in intensity and another scattering peak appeared, which was ascribed to a lamellar assembly of

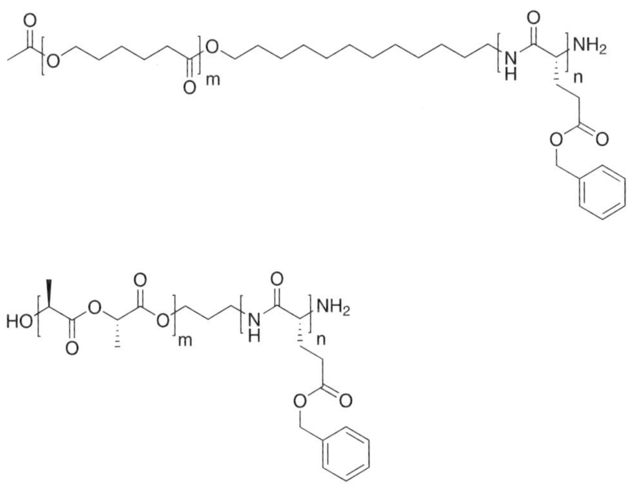

Scheme 7   Chemical structures of poly(ε-caprolactone)-b-poly(γ-benzyl-L-glutamate) (PCL-PBLG) and poly(L-lactic acid)-b-poly(γ-benzyl-L-glutamate) (PLLA-PBLG)

PBLG chains with a β-strand secondary structure. Increasing the temperature to 200 °C not only resulted in melting of PLLA, but also led to a decrease in intensity of the diffraction peaks corresponding to the hexagonally ordered α-helical PBLG segments and an increase in the fraction of PBLG segments that are ordered in a lamellar β-strand fashion.

The structure and properties of an ABA triblock copolymer composed of a polyetherurethaneurea (PEUU) B block with a number-average molecular weight of 15 800 and two PBLG B blocks with a number-average degree of polymerization of 29 were described by Ito and coworkers [74]. DSC thermograms showed a single endotherm located between the glass transition temperature of the PEUU and the 7/2 to 18/5 helix transition of the PBLG segment, suggesting that phase separation did not occur. Platelet adhesion on DMF-cast films of the triblock copolymer was significantly reduced compared to the respective homopolymers and a PEUU/PBLG blend. Thrombus formation on block copolymer films was found to be $\sim$ 50% less compared to a glass surface. However, no significant difference in antithrombogenicity between PEUU, PBLG, their blend and the block copolymer was observed.

Another, very early, study focused on two poly(γ-benzyl-L-glutamate)-b-poly(butadiene-co-acrylonitrile)-b-poly(γ-benzyl-L-glutamate) (PBLG-PBA-PBLG) triblock copolymers [75]. These block copolymers were composed of a PBA block with a number-average molecular weight of 3400 and two PBLG segments containing either 80 or 160 repeat units. TEM micrographs of OsO_4-stained films cast from dioxane, which is a selective solvent for PBLG, revealed a lamellar morphology. The thickness of the PBA layer, as estimated from the TEM images, was 150 Å, while the PBLG layers had thicknesses of 300, respectively 500 Å. The PBLG layer thicknesses pointed towards an end-to-end packing, rather than interdigitation of the PBLG chains. When $CHCl_3$ (which is a non-selective solvent) was used for the preparation of the TEM specimens, the images were more homogeneous and phase separation was less distinct. WAXS experiments showed that the organization of the PBLG blocks was similar to that of the crystalline-like form C modification of PBLG homopolymer. FTIR spectroscopy demonstrated that the PBA block primarily had a *trans*-1,4-butadiene conformation, while the PBLG segments adopted an α-helical secondary structure. The FTIR spectra also indicated that the PBLG helices in dioxane-cast triblock copolymer films are more disordered compared to PBLG helices in $CHCl_3$-cast films. In dynamic mechanical spectra, the side chain transitions of the γ-benzyl-L-glutamate repeat units appeared as a single peak when the triblock copolymers were cast from dioxane and as a double peak when $CHCl_3$ was used as the casting solvent. This suggests that depending on the solvent conditions, the PBA block can influence γ-benzyl-L-glutamate side-chain packing and affect PBLG secondary structure.

Electro- and photoactive peptide–synthetic hybrid triblock copolymers have been prepared using bis(benzyl amine)-terminated poly(9,9-dihexylfluorene-2,7-diyl) (PHF) as a macroinitiator for the ring-opening polymerization of γ-benzyl-L-glutamate N-carboxyanhydride [76]. The electroactive and photoactive properties of the triblock copolymers were similar to those of the PHF homopolymer, indicating that the introduction of the PBLG segments did not interfere with charge injection and transport and other materials properties. The secondary structure of the PBLG segments was studied by means of FTIR spectroscopy. FTIR spectra of $CHCl_3$-cast films of $PBLG_{23}$-PHF_{15}-$PBLG_{23}$ indicated an α-helical secondary structure. The FTIR spectra of the second triblock copolymer, $PBLG_{16}$-PHF_{28}-$PBLG_{16}$, however, contained an additional peak at $1630\ cm^{-1}$, indicating the coexistence of α-helical and β-strand conformations. The thin film morphologies of the triblock copolymers were investigated with AFM using different casting solvents (Fig. 10). When $TFA/CHCl_3$ (30/70 v/v) was used for sample preparation, globular or spherical aggregates with diameters of ~ 32 and ~ 40 nm were observed for $PBLG_{23}$-PHF_{15}-$PBLG_{23}$ and $PBLG_{16}$-PHF_{28}-$PBLG_{16}$, respectively. In this solvent mixture, the PBLG chains have a random coil conformation and the triblock copolymers were proposed to form spherical nanostructures composed of a PHF core and a PBLG shell. When the solvent was changed to $TFA/CHCl_3$ (3/97 v/v), the AFM images revealed parallel fibrillar structures

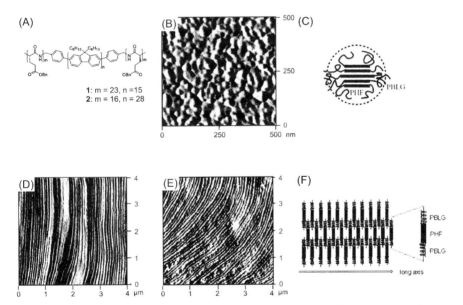

Fig. 10 **A** Chemical structure of PBLG-PHF-PBLG triblock copolymers; **B** AFM image of a thin film of **1** cast from $TFA/CHCl_3$ (30/70 v/v); **C** Schematic representation of the spherical nanostructures that can be observed in **B**; **D,E** AFM images of triblock copolymers **1** (**D**) and **2** (**E**) cast from $TFA/CHCl_3$ (3/97 v/v); **F** Model proposed for the self-assembly of **1** and **2** in the fibrillar structures shown in **D** and **E**. (Reprinted with permission from [76]. Copyright 2004. American Chemical Society)

with widths of 79 ± 25 (PBLG$_{23}$-PHF$_{15}$-PBLG$_{23}$) and 83 ± 17 nm (PBLG$_{16}$-PHF$_{28}$-PBLG$_{16}$) and lengths of $4-10$ μm. Under these solvent conditions, the PBLG segments adopt an α-helical secondary structure. Since the widths of the fibrils were much larger than the extended length of the triblock copolymers, a side-by-side antiparallel stacking was proposed to explain the fibril formation.

9
Conclusions and Outlook

The solid-state structure, organization and properties of peptide–synthetic hybrid block copolymers have been discussed. The most notable difference between peptide–synthetic hybrid block copolymers and their fully synthetic and amorphous analogues is their hierarchical solid-state organization. In contrast to most synthetic amorphous block copolymers, which typically exhibit structural order only over a single length scale, peptide–synthetic hybrid block copolymers can form hierarchically organized nanoscale structures that cover several different length scales. At the smallest length scale, peptide sequences fold into regular secondary structures such as α-helices or β-strands. On the next higher level, these peptide α-helices and β-strands can assemble into hexagonal superstructures and β-sheets, respectively. Finally, phase separation between the peptide and synthetic blocks leads to the formation of ordered domains with the largest characteristic length scales. For a large number of peptide–synthetic hybrid block copolymers lamellar phase-separated morphologies have been observed. These lamellar structures, however, are often found over a much broader range of compositions compared to regular, fully amorphous diblock copolymers. In addition to these more conventional morphologies, structural investigations on peptide–synthetic hybrid block copolymers have also led to the discovery of various novel phase-separated structures, which were not previously known for fully amorphous diblock copolymers. Both observations reflect the fact that the solid-state structure formation of peptide hybrid block copolymers is not solely dictated by phase separation, as is the case for amorphous diblock copolymers, but is also influenced by other factors such as intra- and intermolecular hydrogen bonding and chain conformation. While much of the early interest in peptide–synthetic hybrid block copolymers was driven by their potential use as membrane materials or for the development of antithrombogenic surfaces, more recent studies revealed that these materials can also have interesting mechanical properties [61].

The major drawback of all the block copolymers discussed in this article is that they have been prepared via the conventional amine-initiated NCA polymerization. The polymerization of NCA's under these conditions does not allow very accurate control over polymer chain length, results in rather broad molecular weight distributions and is also not very useful to prepare defined block copolypeptides [77, 78]. It is obvious that these limitations possibly restrict further engineering of the structure and organization of peptide–synthetic hybrid block copolymers and could also hamper the exploration of their full practical potential. Over the past years, however, a number of alternative NCA polymerization strategies have been developed which provide enhanced control over polypeptide chain length and chain length distribution and also allow access to defined block copolypeptides [79–81]. The utility of these synthetic strategies to prepare peptide (hybrid) block copolymers has already been demonstrated. In-depth structural investigations on these materials have yet to be carried out. In a first report, the effect of the polydispersity of the peptide block on the solid-state organization of PS-PZLL block copolymers was discussed [48]. It will be interesting to see to which extent these improved synthetic methods allow enhanced control over the nanoscale structure formation of peptide–synthetic hybrid block copolymers and whether these advances will also influence materials properties and open up new applications.

References

1. Park C, Yoon J, Thomas EL (2003) Polymer 44:6725
2. Hamley IW (2003) Nanotechnology 14:R39
3. Lazzari M, López-Quintela MA (2003) Adv Mater 15:1583
4. Bates FS, Fredrickson GH (1990) Annu Rev Phys Chem 41:525
5. Hamley IW (1998) The physics of block copolymers. Oxford University Press, Oxford
6. Bates FS, Fredrickson GH (1999) Physics Today 52:32
7. Walther M, Finkelmann H (1996) Progr Polym Sci 21:951
8. Mao G, Ober CK (1997) Acta Polym 48:405
9. Muthukumar M, Ober CK, Thomas EL (1997) Science 277:1225
10. Faul CFJ, Antonietti M (2003) Adv Mater 15:673
11. Ikkala O, Ten Brinke G (2004) Chem Commun 2131
12. Klok H-A, Lecommandoux S (2001) Adv Mater 13:1217
13. Lee M, Cho BK, Zin WC (2001) Chem Rev 101:3869
14. Vandermeulen GWM, Klok H-A (2004) Macromol Biosci 4:383
15. Klok H-A (2005) J Polym Sci Part A Polym Chem 43:1
16. Schlaad H, Antonietti M (2003) Eur Phys J E 10:17
17. Deming TJ (2006) Adv Polym Sci 202:1
18. Löwik DWPM, Ayres L, Smeenk JM, van Hest JCM (2006) Adv Polym Sci 202:19
19. Perly B, Douy A, Gallot B (1976) Makromol Chem 177:2569
20. Billot J-P, Douy A, Gallot B (1976) Makromol Chem 177:1889
21. Billot J-P, Douy A, Gallot B (1977) Makromol Chem 178:1641
22. Douy A, Gallot B (1977) Polym Eng Sci 17:523
23. Douy A, Gallot B (1982) Polymer 23:1039
24. Nakajima A, Hayashi T, Kugo K, Shinoda K (1979) Macromolecules 12:840
25. Nakajima A, Kugo K, Hayashi T (1979) Macromolecules 12:844
26. Nakajima A, Kugo K, Hayashi T (1979) Polymer J 11:995

27. McKinnon AJ, Tobolsky AV (1968) J Phys Chem 72:1157
28. Hayashi T, Chen G-W, Nakajima A (1984) Polymer J 16:739
29. Gervais M, Douy A, Gallot B, Erre R (1988) Polymer 29:1779
30. Kugo K, Hayashi T, Nakajima A (1982) Polymer J 14:391
31. Kugo K, Hata Y, Hayashi T, Nakajima A (1982) Polymer J 14:401
32. Kugo K, Murashima M, Hayashi T, Nakajima A (1983) Polymer J 15:267
33. Chen G-W, Hayashi T, Nakajima A (1981) Polymer J 13:433
34. Sato H, Nakajima A, Hayashi T, Chen G-W, Noishiki Y (1985) J Biomed Mater Res 19:1135
35. Yoda R, Komatsuzaki S, Nakanishi E, Hayashi T (1995) Eur Polym J 31:335
36. Yoda R, Komatsuzaki S, Hayashi T (1996) Eur Polym J 32:233
37. Yoda R, Shimoda M, Komatsuzaki S, Hayashi T, Nishi T (1997) Eur Polym J 33:815
38. Yoda R, Komatsuzaki S, Hayashi T (1995) Biomaterials 16:1203
39. Yoda R, Komatsuzaki S, Nakanishi E, Kawaguchi H, Hayashi T (1994) Biomaterials 15:944
40. Babin J, Rodriguez-Hernandez J, Lecommandoux S, Klok H-A, Achard M-F (2005) Faraday Discuss 128:179
41. Gallot B, Douy A, Hayany H, Vigneron C (1983) Polym Sci Technol 23:247
42. Mori A, Ito Y, Sisido M, Imanishi Y (1986) Biomaterials 7:386
43. Imanishi Y, Tanaka M, Bamford CH (1985) Int J Biol Macromol 7:89
44. Janssen K, Van Beylen M, Samyn C, Scherrenberg R, Reynaers H (1990) Makromol Chem 191:2777
45. Janssen K, Van Beylen M, Samyn C, Van Driessche W (1989) Makromol Chem Rapid Commun 10:457
46. Schlaad H, Kukula H, Smarsly B, Antonietti M, Pakula T (2002) Polymer 43:5321
47. Losik M, Kubowicz S, Smarsly B, Schlaad H (2004) Eur Phys J E 15:407
48. Schlaad H, Smarsly B, Losik M (2004) Macromolecules 37:2210
49. Klok H-A, Langenwalter JF, Lecommandoux S (2000) Macromolecules 33:7819
50. Lecommandoux S, Achard M-F, Langenwalter JF, Klok H-A (2001) Macromolecules 34:9100
51. Nishimura T, Sato Y, Yokoyama M, Okuya M, Inoue S, Kataoka K, Okano T, Sakurai Y (1984) Makromol Chem 185:2109
52. Yokoyama M, Nakahashi T, Nishimura T, Maeda M, Inoue S, Kataoka K, Sakurai Y (1986) J Biomed Mater Res 20:867
53. Kugo K, Ohji A, Uno T, Nishino J (1987) Polymer J 19:375
54. Cho C-S, Kim S-W, Komoto T (1990) Makromol Chem 191:981
55. Cho C-S, Kim SU (1988) J Control Rel 7:283
56. Floudas G, Papadopoulos P, Klok H-A, Vandermeulen GWM, Rodríguez-Hernandez J (2003) Macromolecules 36:3673
57. Parras P, Castelletto V, Hamley IW, Klok H-A (2005) Soft Matter 1:284
58. Cho C-S, Jeong Y-I, Kim S-H, Nah J-W, Kubota M, Komoto T (2000) Polymer 41:5185
59. Cho C-S, Jeong Y-I, Kim S-H, Nah J-W, Lee Y-M, Kang I-K, Sung Y-K (1999) Korea Polymer J 7:203
60. Cho CS, Jo B-W, Kwon J-K, Komoto T (1994) Macromol Chem Phys 195:2195
61. Tanaka S, Ogura A, Kaneko T, Murata Y, Akashi M (2004) Macromolecules 37:1370
62. Zhang G, Ma J, Li Y, Wang Y (2003) J Biomater Sci Polymer Edn 14:1389
63. Cho I, Kim J-B, Jung H-J (2003) Polymer 44:5497
64. Cho C-S, Kim S-W, Sung Y-K, Kim K-Y (1988) Makromol Chem 189:1505
65. Cho CS, Song SC, Suh SP, Kim KY, Jang SW, Sung YK (1989) Polymer (Korea) 13:657
66. Hayashi T, Kugo K, Nakajima A (1984) Cont Topics Polym Sci 4:685

67. Kumaki T, Sisido M, Imanishi Y (1985) J Biomed Mater Res 19:785
68. Kang I-K, Ito Y, Sisido M, Imanishi Y (1988) Biomaterials 9:138
69. Kang I-K, Ito Y, Sisido M, Imanishi Y (1988) Biomaterials 9:349
70. Kugo K, Nishioka H, Nishino J (1987) Chemistry Express 2:21
71. Degée P, Dubois P, Jérôme R, Theyssié P (1993) J Polym Sci Part A Polym Chem 31:275
72. Rong G, Deng M, Deng C, Tang Z, Piao L, Chen X, Jing X (2003) Biomacromolecules 4:1800
73. Caillol S, Lecommandoux S, Mingotaud A-F, Schappacher M, Soum A, Bryson N, Meyrueix R (2003) Macromolecules 36:1118
74. Ito Y, Miyashita K, Kashiwagi T, Imanishi Y (1993) Biomat Art Cells & Immob Biotech 21:571
75. Barenberg S, Anderson JM, Geil PH (1981) Int J Biol Macromol 3:82
76. Kong X, Jenekhe SA (2004) Macromolecules 37:8180
77. Kricheldorf HR (1987) α-Aminoacid-N-carboxyanhydrides and related heterocycles. Springer, Berlin Heidelberg New York
78. Deming TJ (2000) J Polym Sci A Polym Chem 38:3011
79. Deming TJ (1997) Nature 390:386
80. Dimitrov I, Schlaad H (2003) Chem Commun 2944
81. Aliferis T, Iatrou H, Hadjichristidis N (2004) Biomacromolecules 5:1653

Adv Polym Sci (2006) 202: 113–153
DOI 10.1007/12_084
© Springer-Verlag Berlin Heidelberg 2006
Published online: 16 March 2006

Drug and Gene Delivery Based on Supramolecular Assembly of PEG-Polypeptide Hybrid Block Copolymers

Kensuke Osada[1] · Kazunori Kataoka[1,2] (✉)

[1]Department of Materials Science and Engineering, Graduate School of Engineering,
The University of Tokyo, 7-3-1 Hongo, Bunkyo-ku, 113-8656 Tokyo, Japan
osada@bmw.t.u-tokyo.ac.jp, kataoka@bmw.t.u-tokyo.ac.jp

[2]Division of Clinical Biotechnology, Center for Disease Biology
and Integrative Medicine, Graduate School of Medicine, The University of Tokyo,
7-3-1 Hongo, Bunkyo-ku, 113-0033 Tokyo, Japan
kataoka@bmw.t.u-tokyo.ac.jp

Abstract Recently, polypeptide hybrid polymers, particularly poly(ethylene glycol) (PEG)-polypeptide block copolymers, have been attracting significant interest for polymeric therapeutics, such as drug and gene delivery systems, utilizing their most relevant feature, that is the formation of micelles with a distinguished core-shell architecture. Of particular interest in the polypeptides is that a variety of functional groups, such as carboxyl groups and amino groups, are available as a side chain, and that they have propensities of low toxicity and biodegradability. The segregated polypeptide core of the micelle embedded in the hydrophilic palisade serves as a reservoir for a variety of drugs as well as of genes with diverse characteristics. The micelles have been developed with various functions, such as biocompatibility, stimuli- and environment-sensitivity, and targetability, aimed at their clinical use. Smart micelles have emerged as promising carriers that enhance the effect of drugs and genes with minimal side effects. In this review, recent advances in drug and gene delivery by polypeptide hybrid micelles, mostly accomplished in our group, are comprehensively described. Focus is placed on the design of PEG-polypeptide hybrid block copolymers, starting from the development of the drug-loading micelle systems to current efforts to establish a gene delivery system with a polyion complex (PIC) micelle, one of the most attractive topics in nanomedicine.

Keywords Block copolymer · Poly(ethylene glycol) · Polyion complex · Polymeric micelle · Non-viral gene vector

Abbreviations

EPR effect	enhanced permeability and retention effect
RES	reticuloendothelial system
PEG	poly(ethylene glycol)
NCA	amino acid N-carboxyanhydride
PLL	poly(L-lysine)
PAsp	poly(α,β-aspartic acid)
PBLA	poly(β-benzyl-L-aspartate)
PGlu	poly(L-glutamic acid)
PEG-PAsp(DPT)	poly(ethylene glycol)-b-poly(3-[(3-aminopropyl)amino]propylaspartamide)
PEG-PAsp(DMAPA)	PEG-poly(3-dimethylamino)propyl aspartamide
PAsp(MPA)	poly[(3-morpholinopropyl) aspartamide]
PEG-PAsp(MPA)-PLL	PEG-b-poly[(3-morpholinopropyl) aspartamide]-b-PLL
PEG-PAMA	PEG-b-poly(2-(dimethylamino)ethyl methacrylate))
PEI	poly(ethyleneimine)
Dox	doxorubicin
PIC micelle	polyion complex micelle
N/P ratio	ratio of [amino group in polycation]/[phosphate group in polynucleic acid]

1
Introduction

Recently, polypeptide hybrid polymers, particularly poly(ethylene glycol) (PEG)-polypeptide block copolymers, have been attracting significant interest for polymer therapeutics, such as drug and gene delivery systems [1–5].

The particular advantages of using polypeptides as hybrid polymers with therapeutic interest are the tailored molecular design through the precise polymerization using the NCA ring opening method, the availability of a variety of functional groups, such as the carboxyl group and amino group, low toxicity, biodegradability and the formation of the characteristic protein folding motif due to the inter- and intramolecular association of peptide chains. A characteristic feature of amphiphilic block copolymers having a large solubility difference between the hydrophilic and hydrophobic segments is the micelle-forming nature in a selective solvent [6–10]. Functional groups, such as the amino group and carboxylic acid, in the PEG-polypeptide hybrid block copolymers may be useful for introducing chemical moieties that modulate the hydrophobicity of the polypeptide blocks as well as conjugating the pharmaceutically active molecules. A series of block copolymers with different functional groups in the side chain may be prepared from the same platform, which is obtained by the polymerization of the appropriate NCA. Systematic control of the structure of the core-forming block leads to a wide variation in drug loading, release, and activation. In contrast to micelles from small surfactant molecules, polymeric micelles are generally more stable and can retain the loaded drug for a prolonged period of time even in a very diluted condition in the body due to an appreciably lower critical micelle concentration (CMC). The micelle formation proceeds through a combination of intermolecular forces, such as hydrophobic interaction [11–20], electrostatic interaction [21–26], metal complexation [27, 28] and hydrogen bonding [29] of the constituent block copolymers. The outer block consists, in many cases, of a PEG block, which will form the shell to surround the core as well as increase the dispersivity of the micelles through steric stabilization. Note that PEG prevents the adsorption of proteins [30, 31] and hence forms a biocompatible shell of polymeric micelle. The size of these micelles is determined by thermodynamic parameters, yet size-control is feasible by variation of the block length. These block copolymer micelles are typically in the size range of several tens of nanometers in diameter with a relatively narrow size distribution, and are therefore similar in size to viruses and lipoproteins. The size and the surface properties of the micelle require careful modulation to achieve longevity in the blood circulation so as to reach the target site in the body [32]. Functionalities at the distal end of the PEG shell contribute to controlling the biocompatibility as well as to incorporate a site-specific property by installing pilot molecules. The inner block can be used to encapsulate or covalently couple active drug molecules, most typically non-polar drugs with a limited solubility in water. The micelle systems based on polymer chemistry may produce future therapeutics [33, 34] and may be able to work as multifunctionalized devices or intelligent nano-devices having combined functions of detection, diagnosis, analysis, and therapy in a single platform.

In this review, recent advances in our research on polymeric micelles based on PEG-peptide hybrid block copolymers for drug and gene delivery are described. The micelle system with antitumor drugs for tumor targeting is reviewed in the following section. The current study to establish a gene delivery system based on a polyion complex (PIC) micelle is then highlighted as one of the most focused topics in nanomedicine.

2
Characteristics of the Micelles Relevant to Drug and Gene Delivery Systems

In order to accomplish an effective drug and gene delivery to the target site through the systemic route, the carriers need to overcome four main barriers as summarized in Fig. 1. First, the carriers need to achieve a long circulation in the blood compartment (Fig. 1a) as a prerequisite for successful targeting. The main obstacles to longevity in the blood circulation of carriers are considered to be the glomerular excretion from the kidney and recognition by the reticuloendothelial system (RES) located in the liver, spleen and lung. Since the threshold molecular weight exists for glomerular filtration (42 000–50 000 for water-soluble synthetic polymers), it can be avoided by increasing the molecular weight of the carriers. Note that the molecular weight of the polymeric micelles is of the order of 10^6 g/mol and they should not excrete through the glomerular route, unless dissociated into unimers. Carriers in the blood circulation may induce non-specific complement activation and opsonization, resulting in the elimination from the blood compartment due to RES recognition. In this regard, the surface modifications of the carriers with biocompatible polymers to provide a stealth character are of crucial importance. Among such biocompatible polymers, PEG is definitely the most commonly used due to its inherent properties of high flexibility, strong hydration, non-toxicity, and weak immunogenecity and has approval from the Food and Drug Administration (FDA). Second, the carriers should be small enough (< 100 nm) to accomplish effective extravasation from the blood compartment to access the target tissue (Fig. 1b). One of the most important reasons for using macromolecular carriers is their preferential accumulation in solid tumors (passive targeting). Such elevated tumor accumulation of macromolecules is currently explained by the microvascular hyperpermeability to circulating macromolecules and the impaired lymphatic drainage in tumor tissues. This phenomenon was termed the "EPR effect" by Matsumura and Maeda [35]. Third, selective uptake into the target cell is necessary for the delivery of drugs with a low permeability through the cellular membranes (Fig. 1c). Since the surface of cells anionically charges due to the existence of sialic acids and proteoglycans, the electrostatic nature of the carrier is an important factor for determining the cellular uptake. Receptor-mediated targeting may be achieved by installing pilot moieties on the surface of the polymeric micelles using end-functionalized block copolymers. Finally, con-

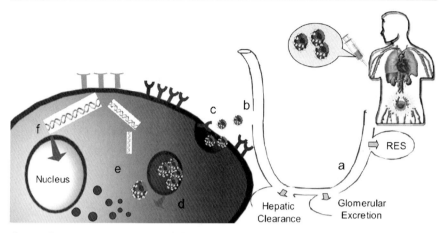

Fig. 1 Schematic representation of the biological distribution of the therapeutic carriers, which are administrated via an intravenous route. The carriers are required to circulate in the blood stream escaping recognition by the RES system, avoiding hepatic clearance and glomerular excretion (a), and extravasate to the tissue through the hyperpermeable region of the capillaries (b). The carriers are taken up by nonspecific or receptor-mediated endocytosis (c), to be transported into the endosome (d). The delivered drugs and genes must be released to the cytosol before lysosomal degradation (e)

trolled intracellular trafficking or nucleus targeting is a matter of importance particularly in gene delivery. Macromolecular carriers (5–100 nm) are taken up by the endocytotic pathway where endosomes with encapsulated carriers are separated from the cell membrane by the process of inward folding. These endosomes (Fig. 1d) have an acidic pH value (pH $\sim$ 5.5), and fuse with lysosomes, where DNA is eventually hydrolyzed by lysosomal enzymes. Thus, a mechanism is needed for a carrier system to safely escape DNA from the endosomal compartment into the cytoplasm.

Carriers, including polymeric micelles, are required to have multi-functionality in order to overcome these barriers and exert their biological action.

3
Stealth Effect of the PEG Shell

PEG is widely used as a key material in a variety of biomedical and pharmaceutical applications due to its properties relevant to in vivo use [36]. One of the primary advantages of using PEG as a shell-forming material for polymeric micelles is its low toxicity. In addition, PEG has long been recognized for its ability to repel proteins at the biological interface [37–39]. The attachment of PEG chains to hydrophobic surfaces substantially reduces protein adsorption due to the hydrophilicity that minimizes the interfacial free energy with water and the high mobility to induce a large exclusion volume.

In PEG-based polymeric micelle systems, the PEG shell contributes to the steric stability of the micelle by physically blocking the flocculation and prevents any non-specific interaction with blood components. The length and density of the PEG chain influence the circulation time and uptake by the RES, with longer chains prolonging the circulation time and decreasing the RES uptake [40]. Thus, encapsulation in the optimized polymeric micelles may be a viable approach to prolonging the circulation time of therapeutic agents.

4
PEG-Polypeptide Hybrid Block Copolymers

Important characteristics of block copolymers for drug delivery include their safety for clinical use and the feasibility of chemical modification for further functionalization. In this regard, PEG-polypeptide hybrid block copolymers, of which the synthetic procedures were mainly established by our group, are quite promising. A series of PEG-polypeptide block copolymers was prepared by the ring opening polymerization of NCA of amino acids with protective groups from the NH_2-terminated PEG. The polymerization proceeded almost quantitatively, and the molecular weight was controlled by the initial monomer/initiator ratio. Block copolymers synthesized by this method have a very narrow molecular weight distribution ($M_w/M_n < 1.1$). The following series of PEG-polypeptide block copolymers was obtained from the corresponding NCA; PEG-PAsp from β-benzyl-L-aspartate (BLA) NCA; PEG-PGlu from γ-benzyl-L-glutamate (BLG) NCA and PEG-PLL from ε-(benzyloxycarbonyl)-L-lysine) (Lys(Z)) NCA followed by deprotection of the protective groups. During the process of debenzylation of PEG-PBLA under alkaline conditions, racemization of the aspartic acid units takes place to form α- and β-aspartate units. The ratio of the α and β units in the PBLA by this treatment is known to be 1 : 3 [41].

Notably, the aminolysis reaction of PBLA quantitatively proceeds even under very mild conditions of room temperature, in which the benzyl ester is replaced by various amino compounds to give a variety of polyaspartamide derivatives. This quantitative aminolysis reaction is quite unique for PBLA, and almost no reaction occurs for PBLG under the same conditions. Presumably, the ester group in the side chain of PBLA may be in the activated form due to the interaction with the amide moieties in the main chain.

Ligand-installed block copolymers for the construction of polymeric micelles useful for active targeting were prepared from heterobifunctional block copolymers possessing different functional groups at the α- and ω-ends. A metal alcoxide of potassium 3,3-diethoxypropanolate was used as an initiator for the ring opening polymerization of ethylene oxide to form heterobifunctional PEG without any side reaction. Subsequent derivatization of the ω-end to the NH_2 group provides the α-acetoxy-ω-amino PEG as a platform polymer for the synthesis of functionalized PEG-polypeptide block copolymers. The acetal moiety at the α-end of PEG can easily be converted into a reactive aldehyde group by gentle treatment with a weak acid solution (pH $\sim$ 2). The aldehyde group at the distal end of the PEG chain is then available for conjugation of targetable ligand molecules such as sugars, peptides, and folate.

Polypeptide hybrid polymers have a controlled biodegradability within the body, in addition to a variety of molecular designs. This characteristic is an important factor for their clinical use. Note that non-degradable polycations accumulate in the nucleus and interact with the gene of the host cells [42]. This effect may cause a long-term toxicity. To overcome this type of toxicity, the molecular design of polycations, degradable in the body to produce non-toxic low molecular weight products, is a crucial issue.

5
Polymeric Micelles for Systemic Cancer Therapy

5.1
Dox-Loaded Micelle

A first generation drug-loaded micelle system developed by our group was the Dox-loaded micelle, which was formed from the Dox-conjugated PEG-PAsp block copolymer (Fig. 2a) [1, 43]. Dox is an antibiotic with a strong tumoricidal activity through the intercalation to DNA strands in the nucleus of tumor cells. In this system, Dox was covalently conjugated to the side chain of the PAsp segment by an amide bond between the carboxylic group in PAsp and the primary amino group of the glycosidyl residue in Dox as shown in Fig. 2b. The substitution ratio of Dox into the PAsp segment was approximately 50%, providing a sufficient hydrophobicity in the PAsp segment to form a stable inner core of polymeric micelles [44] with a diameter of 15–60 nm depending on the composition of the block copolymers and drug contents [45]. The dissociation rate of the micelle estimated in the phosphate buffer saline was on the order of days and was quite slow even in the presence of 50% rabbit serum [46]. In this system, free Dox is physically entrapped into the core of the micelle by a hydrophobic interaction with the conjugated Dox, and the physically entrapped Dox plays a major role in the cytotoxic action of the system. A dimer derivative of Dox molecules via an azomethine bond formation in the micellar core substantially contributes to the micellar stabilization and retention of the loaded drugs. To investigate the biodistribution of the drugs, the physically entrapped Dox in the micelle was radio-labeled with ^{14}C. The Dox micelle showed a remarkably prolonged blood circulation such that 24.6% of the injected dose remained in the blood at 24 h while the free Dox disappeared immediately from the blood (1.6% of the injected dose at 15 min). The

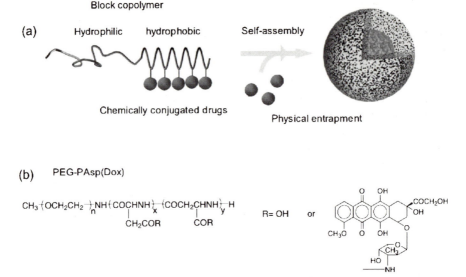

Fig. 2 A first generation of drug-loading micelles. **a** Schematic illustration of the formation of polymeric micelle of Dox-conjugated PEG-PAsp block copolymer. Additional Dox can be physically entrapped in the micelle. **b** Chemical structures of PEG-PAsp block copolymer and Dox

partitioning ratio between the plasma and blood showed that the Dox micelle predominantly existed in the plasma fraction during the circulation (up to 48 h) while the free Dox were distributed to the blood cells. The Dox micelle effectively accumulated in the subcutaneously inoculated tumor (murine colon adenocarcinoma (C-26)) over 24 h, and eventually exhibited a 7.4-fold higher tumor accumulation than the free Dox at 24 h. The effective tumor accumulation of the PEG-PAsp(Dox) micelle, presumably due to the EPR effect, suggests improved therapeutic effects compared to the free Dox. Eventually, PEG-PAsp(Dox) micelles showed a significantly higher in vivo antitumor activity against C-26 than the free Dox. Controlling the composition of the block copolymer and the dose of Dox led to an improved efficacy of the PEG-PAsp(Dox) micelle such that the C-26 tumors completely disappeared [47]. The micellar stability and drug release rate can be controlled by changing the proportion of the chemically conjugated Dox to the PAsp segments and loading amount of the physically entrapped Dox. The optimized PEG-PAsp(Dox) micelle [48] is currently under study in a phase II clinical trial in Japan.

5.2
Cisplatin-Loaded Micelle

Cisplatin (CDDP) is a well-known metal complex exhibiting a wide range of antitumor activity, however, its clinical use is limited due to its significant toxic side effects such as acute nephrotoxicity and chronic neurotoxicity. CDDP shows a rapid distribution over the whole body and high glomerular clearance within 15 min after intravenous injection. Therefore, significant efforts have been devoted to develop a drug delivery system, aimed at increasing the blood circulation period and accumulation in solid tumors [49–55]. However, the unfavorable properties of CDDP have prevented the development of a successful formulation. For instance, CDDP may leak from the liposomes within the blood stream due to the low compatibility between the lipid bilayer and the free CDDP. Hence, many current formulations have utilized a coordination bond between CDDP and polymers or lipids containing carboxylic groups, since two chloride ligands in the leaving group of the Pt(II) atom in CDDP are known to be substituted with a variety of reactive groups depending on the concentration of the chloride ion in the environment [56].

We introduced CDDP into the micelle system where CDDP was complexed with carboxyl groups on PEG-PAsp to form a metal complex micelle (Fig. 3a). The complex spontaneously forms a micelle with a very narrow size distribution having an average diameter of 20 nm [57]. The PEG-PAsp(CDDP) micelles showed an environment responsive drug release behavior. They are stable in distilled water at room temperature, yet in contrast, an exchange between the chloride ion and cisplatin occurred in 150 mM NaCl, resulting in the sustained release of the drug for over 50 h [58].

Biodistribution of the micelle and free CDDP was studied using Lewis lung carcinoma (LLC)-bearing mice. The PEG-PAsp(CDDP) micelle exhibited a time-dependent change in the plasma Pt level. The PEG-PAsp(CDDP) micelle maintained a high plasma Pt level ($\sim 61\%$ of the intravenously injected dose) up to 4–8 h followed by a gradual decrease, while the free CDDP rapidly disappeared from the blood circulation after injection. In contrast to the free CDDP that accumulates in the kidney and causes nephrotoxicity, the PEG-PAsp(CDDP) micelle did not show such a rapid and high Pt accumulation in the kidney up to 15 min. The PEG-PAsp(CDDP) micelle exhibited a 6-fold higher accumulation in the tumor sites compared to the free CDDP at 8 h. Nevertheless, the in vivo antitumor activity was only slightly higher than that of the free CDDP for the same dose (6 mg/kg) [59]. Presumably, the release of CDDP from the PEG-PAsp(CDDP) micelle may not be sufficient to maintain the concentration of the active form of the drugs in the tumor.

Thus, extension of the blood circulation time of the micelles as well as a regulated release rate of the CDDP from the micelle was concluded to be necessary to achieve a more effective anti-tumor activity. This was eventually achieved using PEG-PGlu instead of PEG-PAsp. Here, CDDP was loaded in the micelle in a similar manner to the PEG-PAsp; metal complexation with the ligand substitution reaction between CDDP and the carboxylic group of PEG-PGlu (Fig. 3b) [60]. The formed micelle had a very narrow size distribution with an approximately 30 nm diameter. The PEG-PGlu(CDDP) micelle showed a more sustained release of CDDP (half-value period: > 90 h) than

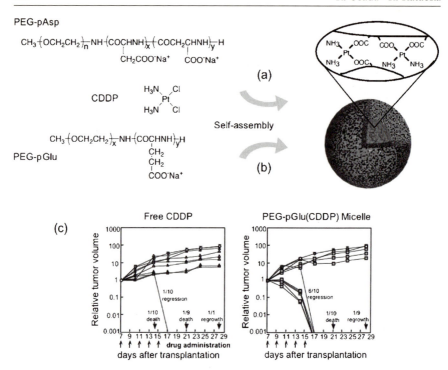

Fig. 3 Polymeric micelle formation of PEG-PAsp/CDDP (**a**), and PEG-PGlu/CDDP (**b**), where carboxylic groups and Pt are linked through coordination bonds. **c** Effect of free CDDP (*left hand side*) and PEG-PGlu(CDDP) micelles (*right hand side*) on the growth of C26 colon adenocarcinoma subcutaneously transplanted in CDF1 mice ($n = 10$). Each drug was administrated by i. v. route five times at 2 day intervals (*arrow*) at the dose of 4 mg/kg CDDP eq. The molecular ratio of CDDP to the block copolymer in the micelles was calculated to be 27, and the dose of the injected polymer was 6.3 mg/kg

the PEG-PAsp(CDDP) micelle (half-value period: ca. 30 h) with a longer induction period (PEG-PGlu(CDDP): > 20 h, PEG-PAsp(CDDP): ca. 10 h) under physiological conditions. A biodistribution assay for the PEG-PGlu(CDDP) micelle revealed a high plasma Pt level with a longer persistent time (11% of the injected dose at 24 h) than the PEG-PAsp(CDDP) micelle (1.5% at 24 h) with a decreased accumulation in the liver and spleen. As a consequence of the longer circulation period, the tumor accumulation of the PEG-PGlu(CDDP) micelle exhibited a 20-fold higher level than that of the free CDDP, indicating a tumor-selective targeting due to the EPR effect. Treatment of the tumor-bearing mice with the PEG-PGlu(CDDP) micelle by intravenous injection achieved complete tumor regression for 5 out of 6 mice with only a minimal body weight loss (within 5% of the initial weight) (Fig. 3c). In contrast, the treatment with the free CDDP with the same drug dose exhibited a tumor regression for only one mouse out of 6 and a significant body weight loss of 20% of the initial weight. The PEG-PGlu(CDDP) micelles are currently undergoing Phase 1 clinical trial (UK).

5.3
Dox-Loaded Micelle with Intracellular pH-Triggered Drug Action

One of the important issues in drug targeting by nanocarrier systems is the selective drug release from the carrier at the target site, thus minimizing the systemic leakage of the loaded drug in the blood stream to ensure safety in their clinical use. In this regard, design of smart polymeric micelles with a stimuli-responsive property is an attractive approach. As for the stimuli to trigger a drug release, difference in proton concentration between intra- and extracellular environment is of interest. Nanocarriers taken into cells via endocytosis are compartmentalized in endosomes (pH $\sim$ 5.5) where the proton concentration increases approximately 100-fold the extracellular condition (pH 7.4). This provides a basis for the design of a smart carrier with a pH-triggered drug releasing mechanism. As a second generation of Dox-loaded micelle, a micelle with an intracellular pH-triggered drug release property was developed recently by our group to improve therapeutic index toward solid tumor.

In a pH-sensitive micellar system, Dox was conjugated to the core-forming PAsp segment of the PEG-PAsp through the hydrazone linker [61] that is stable under physiological conditions but cleavable under the acidic intracellular environments of endosomes and lysosomes (Fig. 4a).

To confirm the acid-sensitive drug release profile, the micelles were incubated under various pH conditions from 7.4 to 3.0. As shown in Fig. 4b, the drugs were released in a time-dependent manner as external pH decreased. Note that no drug release occurred under the physiological condition of pH 7.4 for over 48 h. These results suggest that the micelles may selectively release the loaded drugs under an intracellular acidic condition (pH 5–6) through the cleavage of the hydrazone linkers. Indeed, the intracellular release of conjugated Dox from the micelle was confirmed for multicellular tumor spheroids of a C26 cell line as an in vitro tumor model using a confocal laser scanning microscope [62]. The confocal images showed clear evidence of the intracellular drug release from the micelles and the accumulation of the released drugs into the cell nuclei. In contrast, the drug release from the micelles in the extracellular regions was negligible, being consistent with the result of the model experiments shown in Fig. 4b.

The animal tests revealed that the pH-sensitive micelles showed an effective antitumor activity over a broad range of injection doses to suppress the tumor growth in mice, whereas the toxicity remained extremely low. The micelles were safely injectable up to a 40 mg/kg dose, while three of six mice were completely cured and there was no toxic death among the treated mice. This is in sharp contrast to the free Dox where tumor growth was suppressed

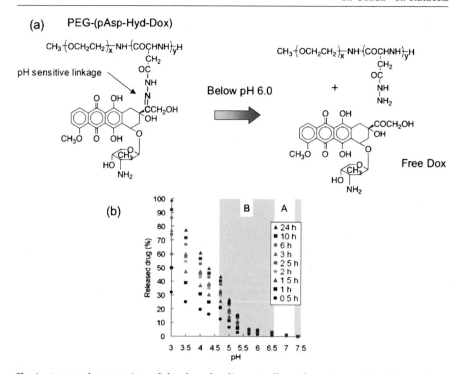

Fig. 4 A second generation of the drug loading micelle with a pH-sensitive drug releasing property. **a** Formation of pH-sensitive polymeric micelles from PEG-(PAsp-Hyd-Dox) block copolymers. Antitumor drugs (Dox), conjugated through acid-labile hydrazone linkers, are released in lower pH conditions. **b** Time- and pH-dependent Dox release profile from the micelles. The micelles selectively release Dox under the pH condition of region B, which corresponds to the intracellular environment. The amount of loaded Dox in the micelles was calculated from the released Dox at pH 3.0 where all of the loaded drugs were assumed to be released from the micelle

with a 10 mg/kg dose, but with a substantial decrease in the body weight due to the toxicity, and toxic death was the result for all of the mice treated with a 15 mg/kg dose of Dox. Namely, the therapeutic efficacy of the micelles was significantly improved over that of the free Dox, which has a narrow therapeutic window between 10 and 15 mg/kg. It should be noted that the design of selective drug releasing systems in the endosomes and lysosomes might escape the drug efflux by P-glycoproteins in multidrug-resistant cells, overcoming a multiple drug resistance in cancer chemotherapy.

Another approach for polymeric micelles using pH-triggered drug release was reported by Lee et al., where an accelerated release of physically incorporated Dox in the micelle was achieved with a decrement of pH [63]. In this report, a pH-sensitive polymeric micelle composed of a mixture of PEG-poly(L-histidine) as a pH-sensitive polybase possessing pK_a values around the physiological pH and biodegradable PEG-poly(L-lactic acid) block copolymers were investigated. The Dox-loaded mixed micelles were stable under the physiological pH condition and destabilized in the pH range of the tumor sites. When the mixed micelles were conjugated with folic acid as the pilot moiety, the micelles were more effective in killing tumor cells due to the accelerated drug release and folate receptor-mediated tumor uptake. Furthermore, the fusogenic activity of poly(L-histidine) in the endosomes facilitated the cytosolic delivery of Dox to achieve an improved cytotoxicity. This approach is also expected to be useful for the in vivo treatment of solid tumors.

6
Polymeric Micelles for Systemic Gene Delivery

The supramolecular drug carriers, such as polymeric micelles, would be the most effective and promising formulation for cancer chemotherapy with an enhanced therapeutic efficacy as well as with a lower toxicity. These characteristics of the micelle system are definitely major advantages for its use in gene delivery systems. Nucleic acid-based drugs have recently attracted increasing attention as a new type of drug, which exert a therapeutic efficacy through the control of the gene expression. Clinical application of these drugs, however, is seriously hampered by their instability under the physiological conditions and the low cellular uptake efficiency due to the large molecular weight and polyanionic nature. Thus, their introduction into appropriate nanocarriers, such as the polymeric micelle, is expected to increase the therapeutic efficacy.

6.1
Gene Therapy

Recently, an in vivo gene therapy mediated by a gene delivery system has been attracting significant attention, since the fast development in biology reveals that quite a number of diseases are caused by gene problems such as mutation. For the treatment of acquired disorders, such as cancer and infectious diseases, effective potential strategies involve not only the introduction of a therapeutic gene, such as the genes for cytokine and antigen, but also the silencing of the expression of abnormal genes in the tissue of the diseased part. In addition to the conventional antisense technology, very recently, RNAi (RNA interference) has become a major mechanism to control the gene expression with therapeutic interest since it has been revealed to have a strong sequence specific gene silencing activity. RNAi is an evolutionarily conserved process in plants and animals by which double-strand RNA induces the sequence-specific degradation of homologous RNA. Synthetic siRNAs can also surrogate for siRNA generated in situ, allowing for target gene knockdown that results in the specific down-regulation of the protein expression in cells. However, a naked DNA or RNA intravenously injected does not lead to sufficient gene regulation because of rapid elimination from the blood stream mainly by DNase and RNase attack in the blood. Therefore, in order to establish a successful gene therapy, it is essential to develop gene vectors, which effectively deliver genes to the target nucleus, thus achieving a high transfection efficiency and persistent transgene expression.

Viral vectors such as retroviruses and adenoviruses have been commonly used in gene delivery since they have high transfection efficiency. However, the use of viral vectors in clinical treatment has problems related to the immune response against viral particles [64, 65], and the possibility of recombination with endogenous viruses as well as oncogene effects [66–68]. Therefore, even though viral vectors have a high efficiency, the establishment of gene delivery systems by non-viral vectors is desirable.

6.2
Non-Viral Gene Vectors

Among non-viral vectors, the lipoplex and polyplex systems [69, 70], in which cationic lipids and polycations, respectively, associate with DNA through an electrostatic interaction, are most widely studied for both in vitro and in vivo transfection. Their assets are that they can carry various size ranges of DNA, ease in manufacturing and mass production, a variety of chemical designs with smart functions, and their surface properties can be readily controlled by changing the charge ratio between the cationic polymer and DNA.

Several lipoplex systems have shown appreciable in vitro transfection activity. Recently, they have been optimized to achieve effective targeting to specific cells as well as the smooth release of the entrapped DNA into the cytoplasm. However, the systems still have unsolved problems, involving stability, non-specific uptake by RES, and cytotoxicity [71]. After an intravenous injection of the lipoplexes, aggregation is immediately induced, then eventually the large aggregates (> 400 nm) are trapped in the lung capillaries and cause an embolism. Thus, the lipoplex systems are still problematic, particularly when used in systemic routes.

An alternative approach for gene vectors is based on polyelectrolytes such as PLL [72] and PEI [73, 74], which form a polyion complex (PIC) with DNA (polyplex) through an electrostatic interaction. Generally, polyplexes require a high N/P ratio for a high stability and efficient transfection activity. Such polyplex systems, containing an excess amount of polycations, may not be suitable for in vivo use, particularly the systemic route, due to their toxicity concerns. The cytotoxicity of polycations, such as PLL and PEI, has been investigated with respect to its correlation with the membrane damaging activity. Polycations can bind to the negatively charged plasma membrane to induce the destabilization. This type of membrane toxicity was quantitatively evaluated by leakage of the lactose dehydrogenase [75, 76]. This effect depends on the time, dose, molecular weight, and the chemical structure of the polycations. A polycation with a higher molecular weight showed a higher membrane-destabilizing activity, while effectively mediating the transfection. Therefore, the molecular weight of the polycations is in a trade off relation between the cytotoxicity and the transfection efficiency. Ogris et al. recently demonstrated that the PEI polyplexes prepared at different N/P feed ratios always gave the identical N/P ratios of 2.5 after purification by size exclusion chromatography, regardless of the initial N/P feed ratio, and concluded that the free PEI mediates the toxicity [77]. Polyplexes prepared at particularly high N/P ratio cause a lethal toxicity mainly due to the embolization of the lung capillary after intravenous administration. Thus, vector design without increasing the N/P ratio is also an important issue for the systemic gene delivery systems. These adverse effects may be overcome by conjugation or by coating of the hydrophilic polymer to the polycation/DNA complex, leading to a reduced non-specific interaction with the blood components, including the erythrocytes. As described here, one of the feasible systems, in this regard, is the polyplex micelles made by the complex formation of DNA with block copolymers composed of a hydrophilic segment and a polycationic segment.

6.3
DNA Condensation Induced by Block Copolymers

In the polyplex systems, DNA is packed in the condensed state. Therefore, it is a very important issue to obtain insight into the mechanism of DNA condensation. DNA undergoes condensation by the process of polyplex formation with various polycations, including spermine, spermidine, PEI, and PLL. Simultaneously, polyplexes often result in the aggregation, and eventually, precipitation due to the charge neutralization. Cationic block copolymers with a hydrophilic segment improve this problem of aggregation due to the formation of polyplex micelles in which the hydrophobic PIC is surrounded by a hydrophilic PEG palisade. In this way, a water-soluble structure with condensed DNA can be obtained without causing aggregation. Note that the micellization would be advantageous not only for use in a gene delivery system but also from the standpoint of the basic study on DNA condensation.

It is an interesting issue to know how the DNA, an inherently rigid macromolecule, condenses into the small and packaged structure in the core of the polyplex micelle. Therefore, the condensation behavior of DNA was investigated to get insight into DNA packaging with the PEG-PLL block copolymer. Note that PLL has been commonly used to study DNA condensation induced by polycations. An S1 nuclease (single-strand specific cleavage enzyme) that cleaves the looped DNA strand was applied to the PEG-PLL/pDNA micelle prepared in a stoichiometric ratio. A surprising digestion behavior by the S1 nuclease was observed as shown in Fig. 5. The S1 nuclease cleaved such a condensed DNA into seven distinct fragments, each being 10/12, 9/12, 8/12,

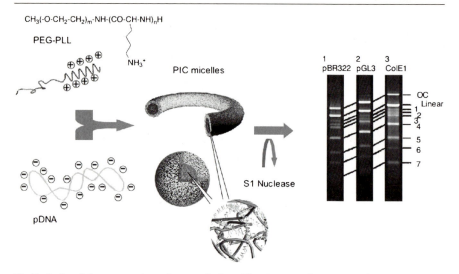

Fig. 5 Ordered fragmentation of pDNA induced by the complexation with PEG-PLL (12–17). S1 nuclease, known to cleave looped DNA strand, is applied to stoichiometrically prepared PIC micelle, in which DNA is condensed in toroid and rod configurations. The condensed DNA is cleaved into seven fragments composed of 10/12, 9/12, 8/12, 6/12, 4/12, 3/12, 2/12 of the original length

6/12, 4/12, 3/12, 2/12 in respective length vs. the original pDNA [78]. More surprising is that the ordered fragmentation occurred in all of the series of pDNAs with different sizes from 2200 bp to 12 000 bp, suggesting that the ordered fragmentation may be related to the inherent propensity of pDNA. This ordered fragmentation suggests that the destabilization of DNA double-strands may arise at ordered sites.

Of interest, the configurations of the polyplex micelle in the stoichiometric ratio are confirmed to be rod and toroid based on AFM observations. On the other hand, the polyplex micelles with a higher charge ratio ($r > 2$) were tightly condensed into a sphere configuration, where no such ordered fragmentation occurred, but pDNA was degraded in a non-specific manner. These observations suggest that there is likely to be a regulated mechanism for DNA folding during the condensation process, and those differences in the manner of DNA condensation in the vectors would affect the transfection activity.

6.4
Polymeric Micelles as Gene Delivery Vector

From the standpoint of designing a PIC micelle-based gene vector [79–85], the major issue is to make the micelle stable enough under physiological conditions, able to dissociate smoothly in the intracellular compartment of the target cells, and facilitating the release of the entrapped DNA. The stability of the micelles is mainly controlled by the charge density and the flexibility of the cationic portion, thus the precise control of the segment length in the block copolymer is of primal importance. Furthermore, in such a design, functions such as specific cell targeting, optimized cellular uptake with specific ligands, and efficient intracellular trafficking should be included.

The formation of nucleic acid loading PIC micelles was first confirmed for a pair of antisense oligo DNA (ODN) and a block copolymer, PEG-PLL [25]. The ODN-incorporated micelle showed a narrow size distribution with a spherical shape. The system was then extended to the DNA with a larger size as the plasmid DNA (pDNA) [79–81]. Even with pDNA, the micelle showed a narrow size distribution with a size range of 100 nm, which was determined by dynamic light scattering (DLS) as shown in Fig. 6a. The nuclease resistance of the PEG-PLL/pDNA PIC micelle was evaluated for the system with varying cationic segment lengths, and the result was compared to that for the native DNA. Degradation of the DNA molecules was evaluated from an increase in the absorbance at 260 nm. As shown in Fig. 6b, the native DNA was non-specifically degraded into fragments immediately after the addition of DNase I, however, no substantial degradation was observed in the micelle system. The micelle with a longer cationic segment showed a higher nuclease resistance, suggesting the more stable complexation with DNA. The stability of pDNA in the physiological media (serum) was also evaluated from the tolerability of the supercoiled pDNA (intact form) against degradation. The fluorescent resonance energy transfer (FRET) between a pair of donor-acceptor fluorescent dyes tagged on a single pDNA molecule correlates with the condensation of pDNA, and was used to estimate the stability of the micelle system in comparison with other systems such as the polyplex (PLL/pDNA) and lipoplex (lipofectamine/pDNA) [86]. It was confirmed that pDNA exerts a high tolerability in a physiological environment through the micellization with the block copolymers; the tolerability in a physiological environment increases with the increasing chain length of the PLL segments in the block copolymer.

Furthermore, the gene transfection efficiency was examined in cultured 293 cells. As shown in Fig. 6c, the transfection efficiency was progressively improved by increasing the length of the PLL segment in the PEG-PLL. Comparing the micelle with the same PEG-PLL composition but at a different mixing charge ratio (r), the micelle with the higher r value revealed a higher gene expression than the one prepared with the stoichiometric system ($r = 1$). The highest transfection was achieved by the PEG-PLL 12–48 (M_w of PEG: 12 000 and unit number of PLL segment: 48)/pDNA micelle with $r = 2$. In terms of the mixing ratio and the length of the PLL segment, the optimum ratio directing high transfection shifts to lower values with increasing length of PLL segments; $r = 3$ in PEG-PLL 12–7, $r = 2.5$ in 12–19, and $r = 2$ in 12–48. This tendency suggests that there seems to be an optimum degree of conden-

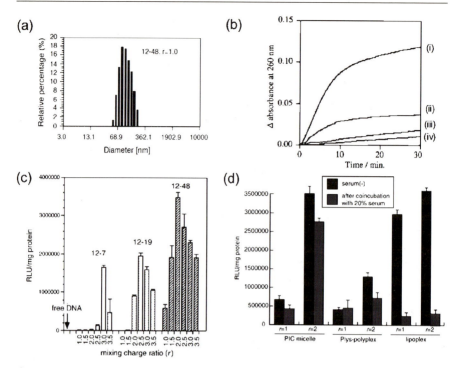

Fig. 6 Basic characterization and transfection activities of PEG-PLL/pDNA micelle. **a** Size distribution of the PIC micelle (PEG-PLL (12–48)/pDNA, charge ratio; 1.0) obtained from histogram analysis of dynamic light scattering (DLS). **b** Nuclease resistance as a function of PLL length evaluated by an increase in absorbance at 260 nm by the addition of DNase I in 10 mM tris-HCL buffer, pH 7.4; (i) Native DNA + 11 U DNase I, (ii) PEG-PLL12-7/DNA complex +110 U DNase I, (iii) PEG-PLL12-19/DNA complex + 110 U DNase I, (iv) PEG-PLL12-42/DNA complex + 110 U DNase I). **c** Charge ratio dependency of transfection activity of PIC micelles with varying compositions (12-7, 12-19, 12-48) against 293 cells ($n = 4$, ± S.D. + 100 µM HCQ). **d** Influence of preincubation with serum on transfection activity of PIC micelles (12-48, $r = 2$), polyplex (PLL/pDNA, $r = 2$), and lipoplex system (LipofectAMINE/pDNA complex). The preincubation of these complexes in 20% serum was done for 30 min prior to transfection ($n = 4$; ± SD). (Fig. 6a,c,d; Reprinted with permission from [87] and Fig. 6b; from [81])

sation for transfection. When the gene transfection efficacy of the cultured cell line of the PEG-PLL 12–48/pDNA ($r = 2$) micelle is compared to those of other non-viral gene vectors, such as the PLL/pDNA polyplex and commercially available lipoplex (LipofectAMINE), the micelle showed a comparable transfection efficiency with the lipoplex and an approximately higher efficacy than the polyplex system (Fig. 6d) [87]. Notably, the PIC micelle retained a sufficient transfection efficiency even after a preincubation with 20% serum for 30 min. In contrast, the lipoplex showed drastic decrease in the efficiency after serum incubation.

To investigate the feasibility of the micelles for in vivo use, biodistribution of the micelles after intravenous injection into mice was studied. The pharmacokinetics was studied for the micelles prepared at varying charge ratios and chain length of the PLL in the PEG-PLL block copolymer [88]. The PIC micelle from PEG-PLL 12–48 at $r = 4$ showed an appreciably long retention time in the blood stream, whereas, the naked pDNA was degraded into small fragments within 5 minutes. The prolonged blood circulation period suggests that DNA is shielded from nuclease attack. The in vivo gene expression (luciferase) was then evaluated for different organs as a function of the charge ratio of the micelles. The expression was only observed in the liver, where the highest expression was achieved for the micelles with a charge ratio of 4 (Fig. 7a). This result is consistent with the tolerability of the micelles within the blood stream. Notably, as shown in Fig. 7b, the gene expression was sustained for 3 days after injection.

The in vivo expression pattern of the micelles was completely different from that of the cationic lipoplex, which showed an expression predominantly in the lung. The lipoplex was probably trapped in the lung capillaries because of their appreciably high positive potential (19–28 mV) to induce a nonspecific interaction with the blood components to form aggregates with larger size. In contrast, the micelle can circulate longer within the blood stream due to the stealth effect mediated by the PEG shell, accordingly, the micelles are not captured in the lung capillary. These properties of the PEG-polypeptide block copolymer micelle system facilitate their future utility in systemic gene therapy.

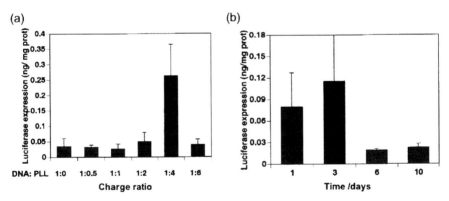

Fig. 7 In vivo gene expression activity of PEG-PLL/pDNA micelle system (12–48) in the liver after supramesenteric injection. PIC micelles were prepared with 50 μg of pGL3 and PEG-PLL block copolymer. After an indicated time from injection, the liver was homogenized and analyzed for luciferase activity. **a** Influence of charge ratios of the micelle 2 days after the injection. Charge ratios are indicated in the figure. **b** Time dependence of gene expression after injection for the micelle with charge ratio of 4. (Fig. 7; Reprinted with permission from [88])

Generally, the PEGylation has been performed to modulate biophysical properties and to decrease inherent toxicity of cationic polyplexes, most generally, by grafting PEG segments onto polycations. Such PEGylated polyplex systems indeed obtain a decreased in vivo toxicity, however, the gene expression efficacy sometimes decreases compared to non-PEGylated polyplex systems [89, 90]. These differences in the gene expression efficacy between the PEG-block-based system and the simple PEG-grafting system suggest that the control of PEG length and density should be very crucial points to appropriately design the non-viral gene vectors with low-toxicity and high gene expression efficacy.

6.5
Polymeric Micelles with Environment Responsive Crosslinking

Remarkable improvements in the protection of DNA from enzyme attack were demonstrated through micellization. Nevertheless, the stability of the pDNA-entrapping micelle is still insufficient in the blood stream for clinical use. The micelle needs to have stable properties during circulation as well as dissociation in the intracellular compartments to release the encapsulated DNA. Several reports have described bioresponsive gene vectors that combine the extracellular stability of DNA by entrapping into polycationic "cages" with disulfide crosslinking. Rapid intracellular release of the DNA is expected upon cleavage of these cages in the reductive environment of the cytoplasm [91–93].

In this regard, smart polymeric micelles were newly designed to have the ability to dissociate in response to chemical stimuli present in the intracellular compartment. The inner core of the micelle was crosslinked through the disulfide bonds, which are cleavable inside the cell [91]. The disulfide bond is known to be stable in the extracellular environment, yet is readily cleaved inside the cell due to the increased concentration of glutathione, the most abundant reducing agent in the cytoplasm. The glutathione concentration is in the millimolar range inside the cell, whereas in the micromolar range in the blood compartment [94]. Certain fractions of the lysine residue of the PEG-PLL block copolymer were replaced by thiol groups, which readily form disulfide crosslinking bonds to form a network structure in the micelle core after DNA complexation (Fig. 9a). Introduction of the thiol groups to the side chains of the lysine segment was accomplished as shown in Fig. 8a, using the heterobifunctional reagent, N-succinimidyl 3-(2-pyridyldithio)propionate (SPDP). This strategy was first examined with ODN [95]. The PIC micelles from the thiolated PEG-PLL and ODN showed no dissociation due to polyion exchange even by adding the excess quantity of the counter polyanion, poly(vinyl sulfate) (PVS), as well as achieving a sufficient colloidal stability due to the PEG shell. The micelle size was approximately 40 nm, which is inherently the same as that of the non-crosslinked mi-

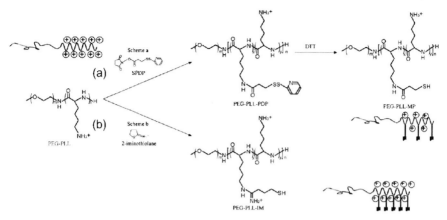

Fig. 8 Schemes for the two methods of thiolation of PEG-PLL. **a** Thiolation by *N*-succinimidyl 3-(2-pyridyldithio)propionate (SPDP) proceeds through the substitution reaction of the ε-amino groups of the lysine residue, resulting in the decreased charge density compensated by the introduction of 3-(2-pyridyldithio)propionyl (PDP) groups via the amide linkage. Treatment of PEG-PLL-PDP with an excess amount of DTT produced the reduced form with flanking 3-mercaptopropionyl groups, PEG-PLL-MP. **b** Thiolation with 2-iminothiolane (Traut's reagent) proceeds through the introduction of cationic imino groups so that the charge density of the PLL segments remained constant (PEG-PLL-IM)

celle. The micelles dissociated to release DNA in the presence of glutathione at a concentration comparable to the intracellular environment. These results indicate that the concept indeed works well at least at the level of the model experiment.

With a special focus on the effect of the charge density of the block copolymer segment as well as the crosslinking density on the transfection efficiency, this system was then extended for pDNA-incorporated micelles [96]. Introduction of thiol groups through substitution of the amino groups of the PLL segment to the amide linkage using SPDP simultaneously decreases the electrostatic association sites between PLL and DNA (Fig. 8a). On the other hand, thiol introduction using Traut's reagent maintains the original charge density, while forming disulfide crosslinking (Fig. 8b). Because of the disulfide crosslinking of the core, both of these thiolated block copolymers formed stable PIC micelles with pDNA with an approximate size of 100 nm. Increased stability due to the core crosslinking was confirmed by the counter polyanion exchange examination. Efficient release of the incorporated pDNA responding to increasing concentrations of the reducing reagent, mimicking the intracellular environment, was only achieved for the system that reacted with SPDP, where the charge decrease through the substitution reaction with SPDP is compensated by the formation of disulfide crosslinking. These distinctive environmental sensitivities were well reflected in the transfection efficiency. The SPDP-reacted system (Fig. 8a), in which the charge density

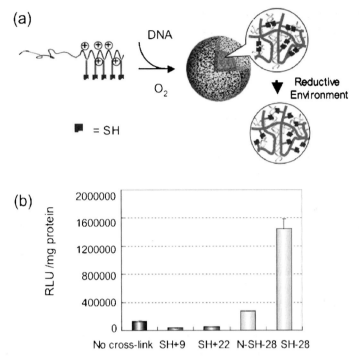

Fig. 9 The crosslinking micelles responding to the intracellular environment. **a** Architecture of a crosslinking micelle. The micelle is stabilized by crosslinking in the extracellular environment, which is readily cleaved in the intracellular environment (reductive environment). **b** Transfection efficiency of the non-crosslinked and crosslinked micelles. SH+9, +22: the micelle with 9 and 28% crosslinking density prepared by the route (b) of Fig. 8, PEG-PLL-IM (the initial charge density is maintained through the reaction). N-SH-28: the micelle from PEG-PLL-PDP prepared by the route (a) of Fig. 8 (28% of amino groups of the PLL were substituted to the amide group), in which no crosslinking occurred due to the presence of a protective group. SH-28: the micelle from PEG-PLL-MP prepared by the route (a) of Fig. 8 (28% of amino groups of the PLL were substituted to the amide group). Crosslinking density is 28%

decreases with an increase in the crosslinking density, revealed an approximately 50 times higher gene transfection than Traut's reagent-reacted system (Fig. 8b), in which the charge density is constant even with the introduction of disulfide crosslinking (Fig. 9b).

From a practical viewpoint, long-term storage of gene carriers is a critical issue. The disulfide crosslinking micelle maintains the original transfection capacity even after freeze-thawing treatment without the use of any protective reagents, while the non-crosslinked micelles showed a significantly lowered transfection efficiency after the same treatment.

Furthermore, the crosslinked micellar vector showed prolonged circulation in the blood compartment, and is currently undergoing an in vivo study, revealing an appreciable gene expression in parenchymal cells of the mouse liver through intravenous injection [97].

6.6
Polymeric Micelles Having the Function of Enhanced Endosome Escape

6.6.1
Escape from the Endosome

For enhancing the transfection efficiency, the delivered DNA must be released from the endosome into the cytosol, but it has to occur before the endosomes fuse with the lysosomes. When a foreign gene is transferred by microinjection or osmotic shock into the cells, the gene is directly delivered into the cytosol, and the gene expression is generally much higher than the usual transfection method using a carrier system because there is no need for the endosome escape process. The importance of endosomal escape is also clearly indicated by the significantly enhanced transfection efficiency of polyplexes in the presence of endosomolytic reagents, such as chloroquine [98, 99], which interferes with the pH lowering in the endosome. A similar enhancement was observed in membrane disruptive peptides, such as oligohistidine [100]. Therefore, if a non-viral vector has the ability to disrupt or fuse with the endosomal membranes, delivered genes can escape from the endosome to the cytosol, and thus, an effective gene expression should occur. Viral envelopes have been known to fuse and destabilize the endosomal and/or lysosomal membrane. For example, influenza virus hemagglutinin (HA) has been extensively investigated and utilized as a pH-sensitive membrane-destabilizing agent. Membrane disruption in the presence of influenza peptide conjugates was demonstrated in a liposome leakage assay, either by electrostatic interaction [99] or by biotin-streptavidin crosslinking [102, 103]. Furthermore, gene transfer using cationic lipid vesicles could be mediated by the fusogenic protein hemagglutinin [104]. Similar to applications with viral peptides, synthesized pH-dependent fusogenic peptides, such as GALA [105], have also been used to promote DNA escape from the endosomes.

An alternative approach to the use of viral components is the design of synthetic pH-sensitive fusogenic lipids [106–108] as well as the use of polycations with a buffering capacity to increase the osmotic pressure in the endosome [109]. One promising strategy to release internalized lipoplexes and polyplexes from the endosome is osmotic endosomal disruption. Behr et al. proposed the hypothesis that PEI by itself has the ability to disrupt the endosomal membrane through the so-called proton sponge effect [73, 74]. Many previous studies in polyplex systems suggest that polycations with a lowered pK_a value such as PEI often show a high transfection efficiency. Note that at pH 5.5–7, PEI has an effective buffering capacity due to its appreciable low apparent $pK_{a,app}$ of 5–6 where the amino groups of PEI are only partially protonated at neutral pH [110]. When the PEI/DNA complex is internalized into the endosome, there occurs a facilitated protonation due to an increase in the proton concentration, eventually raising the ion osmotic pressure of the counter anion in the endosome. This process highlights the substantial role of non-protonated amino groups in the facilitated transport of polyplexes from the endosome to the cytoplasm. Indeed, the degree of protonation of PEI is calculated to increase from 15% to 45% during the pathway from the endosome to the lysosome. The effect of pK_a of the various cationic groups in the polyplexes, having a pK_a value in the range of 7.5 to 8.5, on their transfection efficiency was studied, revealing that the lower the pK_a, the higher the transfection efficiency [111, 112]. However, polycations with lower pK_a values generally have a weak affinity to DNA, and the formed polyplex may be easily dissociated under physiological conditions. On the other hand, polycations with a high pK_a value (> 9.0), such as PLL, form stable polyplexes even at the lower N/P ratio, yet have no buffering capacity under physiological conditions. Putnam et al. introduced buffering units, such as imidazole groups, into the PLL segment to improve the transfection activity based on the proton sponge effect [113]. However, the simple introduction of the buffering units into the polyplex is unlikely to solve the problem of instability under physiological conditions because these buffering units are weak bases and eventually have a low affinity to DNA. In addition, it is known that the protonation of the buffering polycations is facilitated during the complexation with DNA [114], resulting in a loss of the buffering capacity. For the efficient gene vectors, polycations are required to satisfy the conflicting factors of the stabilizing ability and buffering capacity.

6.6.2
Diblock Copolymers With Distinctive pK_a in the Side Chain

Aiming to solve these trade off issues of stability and buffering capacity, we designed new types of block copolymers, which possess both the functions of buffering capacity and high DNA affinity. The block copolymers thus designed have two amino groups with the higher and lower pK_a values in the side chain (Fig. 10a). One amino group with the higher pK_a value acts as the DNA binding portion, while the other amino group with the lower pK_a value acts as the buffer capacity component. The unique feature of this design concept is attributed to the regulated location of the amino groups. The primary amine is located at the very end of the side chain in order to contribute to the polyion complexation with DNA. The secondary amine, located closer to the polymer backbone, is expected to remain in the unprotonated state, presumably because of the lower protonation power and spatial restriction, even though they are combined into a polyion complex. These layouts are expected to enhance the endosomal escape through the buffer capacity, improving the transfection efficiency.

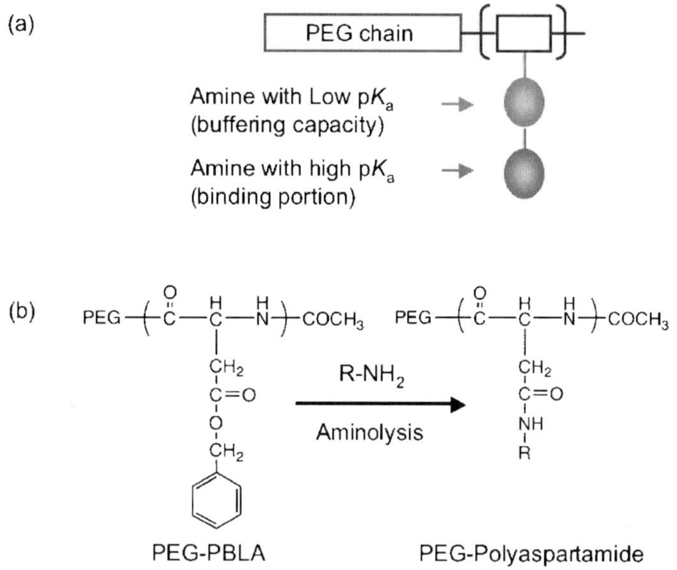

(a)

PEG chain

Amine with Low pK_a
(buffering capacity)   →

Amine with high pK_a
(binding portion)   →

(b)

$$PEG-\left(\!\!\begin{array}{c}O\\ \parallel\\ C\end{array}-\begin{array}{c}H\\ \mid\\ C\\ \mid\end{array}-\begin{array}{c}H\\ \mid\\ N\end{array}\!\!\right)-COCH_3 \qquad PEG-\left(\!\!\begin{array}{c}O\\ \parallel\\ C\end{array}-\begin{array}{c}H\\ \mid\\ C\\ \mid\end{array}-\begin{array}{c}H\\ \mid\\ N\end{array}\!\!\right)-COCH_3$$

R-NH$_2$

Aminolysis

CH$_2$
C=O
O
CH$_2$

CH$_2$
C=O
NH
R

PEG-PBLA   PEG-Polyaspartamide

Fig. 10 Block copolymer with amino group with distinctive pK_a values in the side chain. **a** Design of a new block copolymer with two amino groups with higher and lower pK_a values in each of the monomer units. **b** Schematic representation of the side chain substitution of PEG-PBLA through aminolysis reaction, resulting in the formation of various kinds of PEG-polyaspartamide with different amino groups in the side chain

According to this concept, a library of block copolymers was synthesized through the aminolysis of PBLA (Fig. 10b) in order to search for the best compound for effective transfection. On the basis of both the physicochemical and biological characterizations, PEG-PAsp(DPT) (12–68), as seen in Fig. 11a were found to have a unique feature, in which the propylenediamine units are introduced as pendant groups. The change in the degree of protonation was examined for a model compound of the monomer unit of PAsp(DPT) as shown in Fig. 11b, where a distinctive two-stage protonation process, attributed to the protonation of the primary and secondary amino groups, was observed at 9.9 and 6.4, respectively.

PEG-PAsp(DPT) was then applied to construct a nanocarrier of short interference RNA (siRNA), with the ability to show effective RNA interference (RNAi) properties [115]. RNA is extremely unstable against nuclease attack, and thus the establishment of an efficient delivery system is crucial for promoting the RNAi therapy. The complexation behavior of PEG-PAsp(DPT) with siRNA was examined through gel electrophoresis and EtBr exclusion assay to confirm the formation of stable complexes. The free siRNA disappeared at the N/P ratio > 2 in a gel electrophoresis analysis, which was consistent with the result of the EtBr assay where a substantial fluorescence quenching of EtBr was observed at N/P > 2.

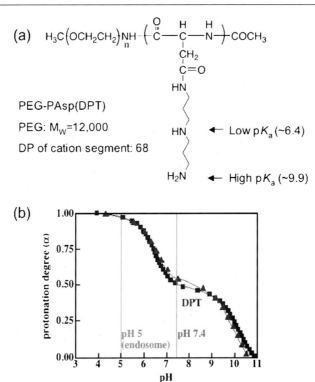

Fig. 11 PEG-PAsp(DPT) block copolymer. **a** Chemical structure of PEG-PAsp(DPT) where dipropylenetriamine is substituted to the benzyl group of PEG-PBLA by aminolysis. **b** The pH-α (α = [protonated amino groups])/[whole amino groups]) curve of BOC-Asp(DPT)-Pr as model compound for PEG-PAsp(DPT). The pK_a values, which are defined as pH at α = 0.25 and 0.75, respectively, were determined to be 9.9 and 6.4

The gene silencing activity of the PEG-PAsp(DPT) complex was investigated. For the gene silencing evaluation, the GL3 luciferase gene was targeted after transfecting two kinds of luciferase pDNAs (pGL3 and pRL) to a Huh-7 cell. As shown in Fig. 12a, each complex system showed a sufficient knockdown of the GL3 luciferase, while neither the naked siRNA nor the PEG-PAsp(DPT) complex with the non-targeting siRNA (mock) showed any knockdown. Notably the gene silencing activity of the PEG-PAsp(DPT) complex was superior especially for the higher N/P ratios (N/P > 10) compared to the other systems, such as PEG-PLL/siRNA and the commercially available RNAiFect/siRNA. At the N/P ratio of 10, the PEG-PAsp(DPT)/siRNA complex showed more than an 80% silencing effect compared to the mock complex. On the other hand, the siRNA/PEG-PAsp(DMAPA) complexes, in which PEG-PAsp(DMAPA) has a dimethyl amino group with a lower pK_a value of 7.9 as the side chain, showed a lower knockdown activity. Presumably, the loosely associated nature of siRNA in the PEG-PAsp(DMAPA) complex, which is

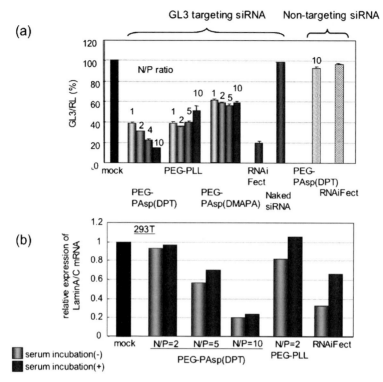

Fig. 12 Evaluation of gene knockdown effect by siRNA complexed with PEG-PAsp(DPT). **a** GL3 luciferase gene knockdown by siRNA complex with various block copolymers having varying charge ratios, commercially available reagent (RNAiFect), and naked siRNA evaluated in HuH-7 cells ($n = 4$; $\pm$ SD). GL3 and RL luciferases were pre-transfected by pGL3 and pRL plasmid DNA complexed with LipofectAMINE. The siRNA complexes (GL3 knockdown) were then applied and evaluated by dual luciferase assay. **b** Endogenous gene (Lamin A/C) knockdown in the presence or absence of 50% serum evaluated in 293T cells

suggested by the EtBr assay, is unfavorable for facilitating the effective intracellular delivery of siRNA in the intact state.

Furthermore, it is notable that the PEG-PAsp(DPT)/siRNA complex showed a significant knockdown of the endogenous gene, the Lamin A/C (cytoskeletal protein), even after a 30 min preincubation in 50% serum (Fig. 12b). The expression was suppressed to the level of 20% of the mock complex, which significantly exceeded the activity of the commercially available reagent of RNAiFect. In contrast, neither the PEG-PLL nor PEG-PAsp(DMAPA) system showed any gene knockdown. The cell viability examined by an MTT assay was more than 75% of the mock cell even after co-incubation with the complex with an N/P > 10, suggesting a lower cytotoxicity. The complex stability under physiological conditions was also examined through incubation in 50% serum at 37 °C prior to the transfection, and it showed a comparable silencing ability even after co-incubation. In contrast, commercially available lipid-based agents such as the RNAiFect system were significantly influenced by serum incubation, probably due to non-specific association with the serum proteins.

The high gene silencing efficiency of the PEG-PAsp(DPT) complex may be attributed to the presence of the secondary amino group with a lower pK_a in the complex that promotes siRNA transport into the cytoplasm by buffering the endosomal cavity, in addition to the increased tolerance against nuclease attack due to complexation. These results, obtained for the engineered block catiomer of PEG-PAsp(DPT), facilitates the clinical use of siRNA for the treatment of various diseases.

6.6.3
Triblock Copolymer Having Tandemly Aligned Segments with Distinctive pK_a Values

A high transfection efficiency was achieved by the diblock copolymer, while such a higher efficiency was obtained for the polyplex with higher N/P ratios. Although the PEG-PAsp(DPT) indeed revealed a lower cytotoxicity, the system still may contain a certain fraction of free polymers that are not involved in the complexation, which may cause cytotoxicity. To improve this problem, an A – B – C triblock copolymer, tandemly aligning two types of polycations with different pK_a values in a single polymer strand was designed [116]. In this triblock copolymer, each segment has its own distinctive roles, expecting a biocompatibility, high transfection efficiency, and stability at the lower N/P ratios. As shown in Fig. 13a, PEG is used as segment A, which acts to form a biocompatible shell when it forms a micelle. Segment B is PAsp(MPA) having an amino group with a low pK_a to induce the buffer effect. Segment C is PLL that acts as the DNA complexing portion.

The degree of polymerization of the PAsp(MPA) and PLL segments in the triblock copolymer were set at 36 and 50, respectively. For comparison, diblock copolymers, PEG-PAsp(MPA) with 39 PAsp(MPA) units and PEG-PLL with 48 PLL units, were also synthesized. The pK_a values of each diblock copolymer were determined to be 6.2 and 9.4, respectively.

At the Lys/nucleotide ratio of 2, the triblock copolymer formed a micelle with a size and zeta potential of 90 nm and + 7 mV. The interaction between the triblock copolymers and pDNA was evaluated and compared to the diblock copolymers using the EtBr exclusion assay. In the case of PEG-PLL, the fluorescence intensity decreased to 20% of the uncondensed naked pDNA at the N/P ratio of two. In contrast, PEG-PAsp(MPA), which has a cationic segment with a lower pK_a value, maintained a relatively high fluorescence (> 90%) over a wide range of N/P ratios, suggesting that PEG-PAsp(MPA) lacks the capacity to condense pDNA based on the EtBr assay result. On the other hand, PEG-PAsp(MPA)-PLL exhibited a 80% decrease in fluorescence

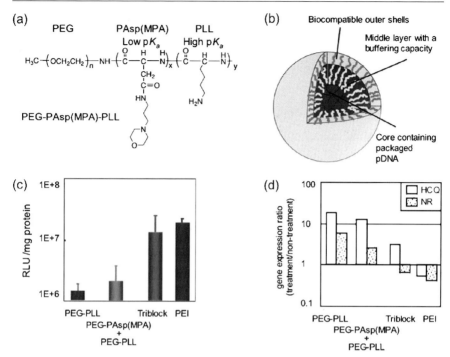

Fig. 13 Panel for triblock copolymer PEG-PAsp(MPA)-PLL system. **a** Chemical structure of PEG-PAsp(MPA)-PLL. **b** Schematic illustration hypothesizing a three-layered micelle formed from the triblock copolymer and pDNA with spatially regulated structure. **c** In vitro transfection of the luciferase gene to HeLa cells by the micelles from di- or triblock copolymers and polyplex with PEI. The micelles were prepared at a Lys/nucleotide ratio of 2. HeLa cells were incubated with each micelle in a medium containing 10% serum for 24 h, followed by additional 24 h incubation without micelles. **d** The effects of HCQ and NR on the transfection efficiency of the micelles and polyplex. The PEI polyplex was prepared at a N/P ratio of 10. (Fig. 13d; Reprinted with permission from [116])

at the N/P ratio of 3. Notably, the fluorescence curve of the triblock system was almost identical to that of the PEG-PLL system when the N/P ratio was converted to the lysine/nucleotide unit ratio. Therefore, in the micelle of the triblock copolymer, the PLL segment may predominantly contribute to the condensation of the pDNA. This assumption was confirmed by the ^{1}H-NMR measurement in deuterated PBS buffer (pD = 7.4, 150 mM NaCl), in which the chemical shifts assigned to the PLL segments completely disappeared but those assigned to the PAsp(MPA) segments still remained in the spectrum. These results are consistent with the original idea that the PEG-PAsp(MPA)-PLL/pDNA complex may form a three-layered micelle as illustrated in Fig. 13b.

Next, the transfection activity of the system was evaluated, expecting a higher efficiency due to the enhanced endosome escape. Obviously, as shown in Fig. 13c, the triblock copolymer, with a lysine/nucleotide ratio of 2, revealed a one order of magnitude higher transfection compared to the PEG-PLL block copolymer. This transfection efficiency is comparable to that of the PEI/pDNA complex, but with a remarkably lower cytotoxicity. On the other hand, the (PEG-PAsp(MPA) + PEG-PLL)/pDNA, where the contents and the repeating units of the PAsp(MPA) and PLL segments were nearly equal to the triblock copolymer, had almost the same transfection efficiency as that of PEG-PLL. The PEG-PAsp(MPA)/pDNA system showed no transfection activity over a wide range of N/P ratios. These results indicate the importance of aligning in a tandem manner, the two types of polycations with adequate pK_a values in a single polymer strand. To confirm that the high transfection activity of the triblock copolymer system was attributed to the enhanced buffer effect, the transfection activity was compared with and without hydroxychloroquine (HCQ) and nigericin (NR). HCQ is a reagent known to increase the transfection activity of polyplexes without any functions to facilitate endosome escape, whereas NR decreases the transfection activity of polyplexes showing a proton sponge effect. As shown in Fig. 13d, the triblock copolymer system showed less effect by HCQ on enhancing the transfection activity compared to the PEG-PLL/pDNA, while it showed an appreciable decrease in the transfection efficiency in the presence of NR. These results suggest that the enhanced transfection activity of the triblock copolymer may be attributed to the proton sponge effect of the intermediate PAsp(MPA) layer of the micelle.

The tandem alignment of two types of polycations with distinctive pK_a values might allow the preferential interaction of the high pK_a block with pDNA, preventing the low pK_a block from facilitated protonation during the complexation with pDNA. Thus, the three-layered system enables one to enhance the transfection activity under the condition where free or loosely associated polycations are assumed to be minimal, providing a new design for the vector useful in systemic gene delivery.

6.7
Organic–Inorganic Hybrid Micelle

Another strategy to enhance the transfection activity is to control the release of the DNA from the micelle responding to a change in the biological microenvironment, such as ion concentration. Selective release of the complexed DNA in the intracellular compartment can be achieved based on the different calcium ion concentrations between the intra- and extracellular environments.

So far, a coprecipitate of calcium phosphate and DNA (CaP/DNA) has been widely used for transfection and for gene silencing by the ODN because of its biocompatibility with CaP [117–119]. However, crystal growth of CaP is uncontrollably fast so that the transfection activity steeply decreases soon after the initial mixing of the calcium and phosphate solution due to the generation of large precipitates.

A new methodology to control the CaP crystal growth was developed using a PEG-PAsp [120]. The block copolymer regulates crystal growth presumably through the adsorption of the PAsp segment onto the crystal surface to form the PEG palisade, thus decreasing the interfacial free energy. Eventually, this method allows one to obtain an inorganic–organic hybrid micelle with a core of the CaP crystal and pDNA surrounded by a PEG shell (Fig. 14a). The particle size measured by DLS is around 100 nm with a significantly narrow size distribution. Cytotoxicity of the hybrid micelle was evaluated by the MTT assay, showing that the hybrid micelles have an essentially non-toxic nature. This is the great advantage of this system. The difference in the intra- and extracellular concentrations of the calcium and phosphate ions, which influence the dissolution behavior of the CaP crystals, is the basis for triggering DNA

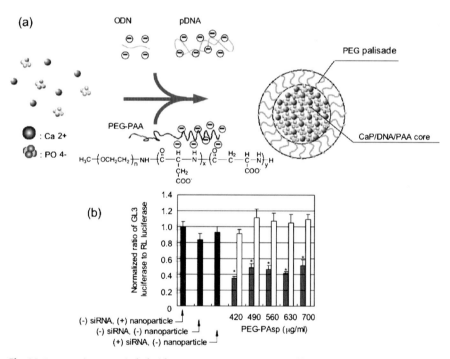

Fig. 14 Inorganic–organic hybrid CaP/PEG-PAsp/DNA micelle system. **a** Schematic representation of organic–inorganic hybrid micelle formation. **b** Biological activities of siRNA incorporated hybrid nanoparticles formed at various PEG-PAsp concentrations. Ratios of GL3 luciferase to RL luciferase were normalized to cells treated with nanoparticles formed without siRNA. *Grey* and *white bars* indicate the ratios of GL3 to RL in the presence of the nanoparticles loading siRNA targeting GL3 luciferase and non-silencing siRNA (mock) used as a control, respectively. *Stars* indicate significant difference, with $p < 0.01$ (*) ($n = 6$, $\pm$ SEM)

release from the hybrid micelle. The extracellular concentration of free calcium ions is approximately 2 mM, while that in the intracellular fluid abruptly decreases to the order of 100 nM [121]. On the other hand, the concentration of the phosphate ion increases from 1 to 40–70 mM through an environment change from the extracellular to intracellular compartment [122]. Dissociation kinetics of the hybrid nanoparticles under the condition mimicking the intra- and extracellular environments showed the gradual dissociation of the nanoparticles when placed in the intracellular conditions [123]. This difference in ionic concentration led to the dissociation of the CaP crystal, which induces the hybrid nanoparticle to dissociate, resulting in the selective release of the loaded DNA. Laser confocal microscopy observation for the intracellular distribution of the nanoparticles suggested that DNA was successfully released from the nanoparticles into the cytoplasmic compartment. The biological significance of these nanoparticles carrying pDNA was then evaluated. The transfection activity of the hybrid nanoparticle was evaluated for 293 cells and HeLa cells for comparison with a conventional pDNA/CaP particle. The expression of the luciferase gene by the hybrid nanoparticle is seven times higher (in the absence of fetal calf serum) and four times higher (in the presence of the serum) than that of the conventional system [124].

Furthermore, the RNAi activities of siRNA were examined using the prepared hybrid nanoparticles (Fig. 14b). The inhibitory activity was evaluated from the relative silencing of the pGL-3-luciferase expression against the pRL-TK expression used as the internal standard [123]. While naked siRNA, empty nanoparticles, and non-silencing siRNA loaded nanoparticles (mock) used as a control showed a negligible silencing effect, appreciable silencing of the GL3 luciferase gene expression (up to 60%) was observed for the siRNA-loaded hybrid nanoparticles prepared over the polymer concentration range from 420 to 700 µg/ml.

These results demonstrate the feasibility of the in vivo use of the organic–inorganic hybrid nanoparticles especially due to their extremely low cytotoxicity. Furthermore, this system might have possible application as a carrier for versatile compounds, such as anticancer drugs and proteins, because of the high binding affinity of CaP to these compounds.

6.8
Active Targeting

6.8.1
Ligand Conjugation to Polymeric Micelles

Accumulation at the tumor sites of the systematically injected micelles described so far is achieved by passive targeting mainly by the EPR effect. To increase the delivery efficiency and to decrease the side effects, the concept of active targeting, which is mediated by ligands binding to the receptors over-expressed in the targeted cells, is attractive. To develop gene vectors with the ability to distinguish between the target and non-target tissue, a general strategy is to modify them with cell-binding ligands that recognize receptors specifically expressed on the target cells.

In hepatocytes, a large number of cell-surface receptors that bind and subsequently internalize the asialoglycoprotein (ASGP) are extensively expressed. Thus, in order to achieve cell-specific gene transfection toward liver parenchymal cells, a galactose moiety, which is recognized by the ASGP receptors, may be introduced onto the surface of the vectors. Systemic delivery of the galactosylated vectors to the hepatocytes has been applied based on ASGP receptor targeting [126–129].

Macrophages are an important target for the gene therapy of diseases, such as Gaucher's disease and human immunodeficiency virus (HIV) infection. In these cases, mannose ligands are available because large numbers of mannose receptors are expressed on the surface of the macrophages. Mannnosylated vectors also have potential applications in DNA vaccine therapy, because antigen-coded pDNA must be efficiently transfected into dendritic cells, which express a large number of mannose receptors. Transferrin, an iron-binding glycoprotein, is a well-studied ligand for tumor targeting. In rapidly dividing cells, such as malignant cells, expression of the transferrin-receptors on their surfaces is elevated due to an increased cellular demand for iron. The folate receptor is overexpressed in a large fraction of human tumors, but is only minimally distributed in normal tissue, indicating that this receptor is also available for tumor-targeting therapy.

The versatile designs and engineering of block copolymers enable one to prepare the polymeric micelles with a targetability to specific tissues. Indeed, micelles mounting sugar [130], peptides [131, 132], and monoclonal antibodies [133] on their surfaces have so far been reported.

6.8.2
Polymeric Micelles with Pilot Molecules

The efficacy of ligands were evaluated in a lactose-installed PEG-PAMA block copolymer micelle encapsulating pDNA (Fig. 15a) [134, 135]. The transfection efficiency was evaluated against an HepG2 cell (hepatoma) possessing an abundant ASGP receptor (150 000 binding sites/cell on their surface), which recognizes compounds bearing the terminal β-D-galactose residue. Figure 15b shows the time dependent gene transfection for the complex with or without the lactose ligand (N/P: 6.25) and the LipofectAMINE (charge ratio: 5) in the presence of 5% FBS. The lactose-micelle achieved a significantly higher transfection efficiency than the non-ligand micelle, which may be attributed to the receptor-mediated endocytosis. In order to confirm the receptor-mediated mechanism, a competitive assay using asialofetuin (ASF), a natural ligand against ASGP receptors, was performed. ASF should work

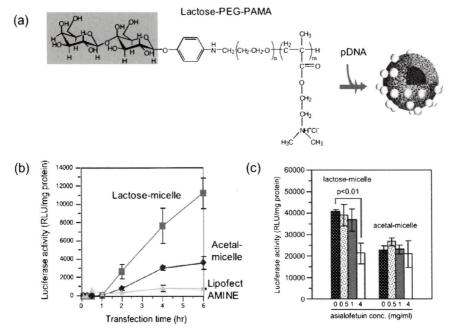

Fig. 15 Panel for the micelle with targeting moiety. **a** Chemical structure of lactosy-lated PEG-PAMA and its micelle formation with pDNA. **b** Effect of transfection time on gene expression. HepG2 cells were transfected with acetal- or lactose-micelles prepared at N/P = 6.25 in the medium (DMEM + 5% FBS) containing 100 μM HCQ. Transfection with LipofectAMINE was done in the same medium without HCQ (± SEM, $n = 4$). **c** Inhibitory effect of asialofetuin (ASF) on gene transfer to HepG2 cells co-incubated with the micelles with or without ligand moiety. The transfection time was fixed to 6 h. (± SEM, $n = 4$) (Fig. 15c; Reprinted with permission from [135])

as an inhibitor of the lactose-micelle, unless the receptor-mediated mechanism does not play a substantial role. As shown in Fig. 15c, the transfection efficiency decreased with an increasing ASF concentration. Especially, in the presence of the ASF with a concentration of 4 mg/ml, the transfection efficiency decreases to half the value of the transfection without an inhibitor. On the other hand, the transfection efficiency of acetal-micelle was not affected by the presence of ASF. This result indicates ASGP-receptor-mediated endocytosis to be a major pathway for the cellular uptake of the lactosylated micelle.

The effect of the lactose ligand was reconfirmed in the micelle system composed of the PEG-siRNA conjugate where the various functions developed so far were integrated [136]. As shown in Fig. 16a, the lactose was installed in the α end of the biocompatible PEG segment. An siRNA was directly conjugated via an acid labile β-thiopropionate linkage at the ω end of the α-lactosylated PEG, which was readily cleaved at the pH corresponding to that of the intra-

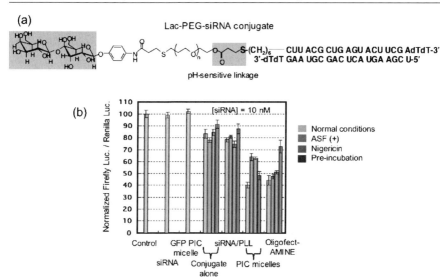

Fig. 16 Panel for ligand-attached PEG-siRNA conjugate through the acid-labile linkage system. **a** Chemical structure of Lac-PEG-siRNA conjugate, which is readily cleaved at the pH corresponding to that of the endosome (pH 5.5). **b** Evaluation of RNAi activities against the firefly luciferase coded gene in cultured HuH-7 cells under various conditions. Normalized ratios between the firefly luciferase activity (firefly luc.) and the renilla luciferase activity (renilla luc.) are shown as *ordinate* ($n = 3$, $\pm$ SD)

cellular endosomal compartment (pH $\sim$ 5.5). Apparently, the shielding layer of PEG hampers intracellular activity of siRNA, and the micelle may be required to detach the PEG shield after its entry into the target cell by endocytosis. The exposed polyplex would then cause an efficient destabilization of the endosomal membrane. For this purpose, bioresponsive PEG-siRNA conjugates with a pH-labile linkage, which may detach the PEG shield in the acidic milieu of the endosomes, was designed.

By mixing the lactosylated PEG-siRNA conjugate and PLL (degree of polymerization, 40) as a cationic polyelectrolyte in a stoichiometric ratio, a spherical core-shell micelle was spontaneously formed. The micelle formation was also confirmed by TEM observations. The gene silencing activity of the system was evaluated through a dual luciferase reporter gene assay in cultured HuH-7 cells (human hepatoma cells) possessing ASGP receptors. The Lac-PEG-siRNA/PLL micelle, PEG-siRNA/PLL micelle, Lac-PEG-siRNA conjugate alone, siRNA/PLL polyplex, naked siRNA, and commercially available lipoplex of oligofectAMINE were compared (Fig. 16b). Even in the presence of 10% FBS, the ligand installed micelle with the N/P ratio of 1 showed a RNAi activity in a dose-dependent manner. The Lac-PEG-siRNA/PLL micelle showed a 100 times more effective RNAi activity in a 50% inhibitory concentration (IC_{50}) compared to the Lac-PEG-siRNA conjugate alone. The siRNA alone showed no RNAi effect probably due to rapid degradation in the serum. The polyplex, siRNA/PLL (N/P = 1) showed a significantly lower RNAi effect compared to the micelle presumably due to the aggregation in the charge neutralized condition and non-specific association between the serum proteins. An oligofectAMINE polyplex showed a remarkable silencing ability, but a significantly reduced ability (56% → 27% inhibition) once it was preincubated in 50% serum for 30 min. This is in sharp contrast to the PIC micelle. To investigate the efficacy of the ligand conjugation, ASF, the inhibitor for the ASGP receptor-mediated endocytosis, was added to the cultured medium. Consequently, the RNAi activity was significantly reduced for the lactosylated PIC micelle (60% → 36% inhibition). As for the control experiment, the same evaluation was done in NIH3T3 cells (mouse fibroblast), which have no ASGP receptor. Eventually, no effect by the ASF addition was observed. These results obviously indicated that the enhanced RNAi activity for the lactose-installed micelle system is attributed to the increased receptor-mediated endocytosis. The contribution of the acid-labile linkage was also studied by adding NR, an inhibitor for the endosomal acidification. The RNAi activity was drastically reduced for the PIC micelle (60% → 37% inhibition) while no effect was observed for the Lac-PEG-siRNA conjugate alone, the siRNA/PLL polyplex, and the lipoplex. These results indicate that after internalization by the receptor-mediated uptake, the linker was cleaved in a manner synchronized with the pH decrease in the endosomal compartment. The release of hundreds of free PEG strands increased the colloidal osmotic pressure as well as destabilizing the endosome membrane by the hydrophobic siRNA/PLL complex. This may induce swelling and rupture of the endosome, facilitating the transport of siRNA into the cytoplasm.

7
Summary

The most relevant feature of using block copolymer micelles for drug and gene delivery is their distinguished core-shell architecture. Among a variety of block copolymers, the PEG-polypeptide hybrid block copolymers have significant advantages due to their unique properties including the availability of various chemical modifications as well as a reduced toxicity. By selecting appropriate polypeptide segments in the constituent block copolymer, the formed micelles have properties and functions relevant for the delivery system. The micelles exhibited several preferable propensities such as a high colloidal stability, reduced interaction with biological components, and prolonged circulation in the blood, and thus, are recognized as promising nanocarriers for enhancing the efficacy of drugs. Indeed, several micelle systems loading antitumor drugs are currently undergoing clinical trials. Recent research has focused on the polymeric micelles for drug and gene delivery with the smart functions such as targetability to specific tissues and responsivity to chemical and physical stimuli. Such smart micelles enhance the efficacy of the loaded drugs as well as minimize side effects, providing a new regime to enhance the efficacy of drug and gene therapy in a safe and secure manner. The development of smart polymeric micelles is a subject that is attracting growing attention and will be continuously studied in the next decade.

References

1. Kataoka K, Kwon GS, Yokoyama M, Okano T, Sakurai Y (1993) J Contr Release 24:119
2. Duncan R (2003) Nat Rev Drug Discovery 2:347
3. Haag R (2004) Angew Chem Int Ed 43:278
4. Lavasanifar A, Samuel J, Kwon GS (2002) Adv Drug Deliv Rev 54:169
5. Kakizawa Y, Kataoka K (2002) Adv Drug Delivery Rev 54:203
6. Moffit M, Khougaz K, Eisenberg A (1996) Acc Chem Res 29:95
7. Tuzar Z, Kratochvil P (1976) Adv Colloid Interface Sci 6:201
8. Munk P, Prochazka K, Tuzar Z, Webber SE (1998) CHEMTECH 28:20
9. Talingying MR, Munk P, Webber SE, Tuzar Z (1999) Macromolecules 32:1593
10. Kataoka K, Harada A, Nagasaki Y (2001) Adv Drug Delivery Rev 47:113
11. Bader H, Ringsdolf H, Schmidt B (1984) Ang Makromol Chem 123/124:457
12. Yokoyama M, Inoue S, Kataoka K, Yui N, Okano T, Sakurai Y (1989) Makromol Chem 190:2041
13. Yokoyama M, Miyauchi M, Yamada N, Okano T, Sakurai Y, Kataoka K, Inoue S (1990) J Contr Release 11:269
14. Kabanov AV, Chrkhonin VP, Yu V, Alakhov V, Batrakova EV, Levedev AS, Melik-Nubarov NS, Arzhakov SA, Levashov AV, Morzov GV, Severn ES, Kabanov AV (1989) FEBS LETT 258:343
15. Kwon GS, Naito M, Yokoyama M, Okano T, Sakirai Y, Kataoka K (1993) Langmuir 9:945
16. Bazile D, Prud'homme C, Bassoulett M-T, Marland M, Spenlehauer G, Vrillard M, Stealyh M (1995) J Pharm Sci 84:493
17. Hagan SA, Coombes AGA, Garnett MC, Dunn SE, Davis MC, Illum L, Davis SS, Harding SE, Purkiss S, Gellert PR (1996) Langmuir 12:2153
18. Gref R, Minamitake Y, Peracchia MT, Trubetskoy V, Torchilin V, Langer R (1994) Science 263:1600
19. Zhang X, Jackson JK, Burt HM (1996) Int J Pharm 132:195
20. Yasugi K, Nagasaki Y, Kato M, Kataoka K (1999) J Contr Release 62:89
21. Harada A, Kataoka K (1995) Macromolecules 28:5294
22. Harada A, Kataoka K (1999) Science 283:65
23. Kabanov AV, Bronich TK, Kabanov VA, Yu K, Eisenberg A (1996) Macromolecules 29:6797
24. Bronich TK, Kabanov AV, Kabanov VA, Yu K, Eisenberg A (1997) Macromolecules 30:3519
25. Kataoka K, Togawa H, Harada A, Yasugi K, Matsumoto T, Katayose S (1996) Macromolecules 29:8556
26. Harada A, Kataoka K (1998) Macromolecules 31:288
27. Yokoyama M, Okano T, Sakurai Y, Suwa S, Kataoka K (1996) J Contr Release 39:351

28. Nishiyama N, Yokoyama M, Aoyagi T, Okano T, Sakurai Y, Kataoka K (1999) Langmuir 15:377
29. Kataoka K, Ishihara A, Harada A, Miyazaki H (1998) Macromolecules 31:6071
30. Jeon SI, Lee JH, Andrade JD, De Gennes PG (1991) J Colloid Interface Sci 142:149
31. Otsuka H, Nagasaki Y, Kataoka K (2001) Curr Opin Colloid Interface Sci 6:3
32. Kataoka K, Kwon GS, Yokoyama M, Okano T, Sakurai Y (1993) J Contr Release 24:119
33. Thanou M, Duncan R (2003) Curr Opin Invest Drugs 4:701
34. Moghimi SM, Hunter AC, Murray JC (2005) FASEB J 19:311
35. Matsumura Y, Maeda H (1686) Cancer Res 46:6387
36. Trubetskoy VS, Torchilin VP (1995) Adv Drug Deliv Rev 16:311
37. Wang P, Tan KL, Kang ET (2000) J Biomater Sci Polym Ed 11:169
38. Lee JH, Lee HB, Andrade JD (1995) Prog Polym Sci 20:1043
39. Joen SI, Lee JH, Andrade JD, de Gennes PG (1991) J Colloid Interface Sci 142:149
40. Kwon G, Suwa S, Yokoyama M, Okano T, Sakurai Y, Kataoka K (1994) J Contr Release 29:17
41. Wolk SK, Swift G, Paik YH, Yocom KM, Smith RL, Simon ES (1994) Macromolecules 27:7613
42. Godbey WT, Wu KK, Mikos AG (1999) Proc Natl Acad Sci USA 96:5177
43. Yokoyma M, Okano T, Sakurai Y, Ekimoto H, Shibazaki C, Kataoka K (1991) Cancer Res 51:3229
44. Yokoyma M, Okano T, Sakurai Y, Kataoka K (1994) J Contr Release 32:269
45. Fukushima S, Machida M, Akutsu T, Shimizu K, Tanaka S, Okamoto K, Machida H, Yokoyma M, Okano T, Sakurai Y, Kataoka K (1999) Colloids and Surfaces B: Biointerfaces 16:227
46. Yokoyma M, Sugiyama T, Okano T, Sakurai Y, Naito M, Kataoka K (1993) Pharm Res 10:895
47. Yokoyma M, Kwon GS, Okano T, Sakurai Y, Ekimoto H, Okamoto K, Mashiba H, Seto T, Kataoka K (1993) Drug Deliv 1:11
48. Matsumura Y, Hamaguchi T, Ura T, Muro K, Yamada Y, Shimada Y, Shirao K, Okusaka T, Ueno H, Ikeda M, Watanabe N (2004) Br J Cancer 91:1775
49. Newman MS, Colbern GT, Working PK, Engers C, Amantea MA (1999) Cancer Chemother Pharmacol 43:1
50. Avichechter D, Schechter B, Arnon R (1998) React Funct Polym 36:59
51. Bogdanov A Jr, Wright SC, Marecos EM, Bogdanova A, Martin C, Petherick P, Weissleder R (1997) J Drug Targeting 4:321
52. Perez-Soler R, Han I, Al-Baker S, Khokhar AR (1994) Cancer Chemother Pharmacol 33:378
53. Ohya Y, Masunaga T, Baba T, Ouchi T (1996) J Biomater Sci Polymer Edn 7:1085
54. Ferruti P, Ranucci E, Trotta F, Gianasi E, Evagarou EG, Wasil M, Wilson G, Duncan R (1999) Macromol Chem Phys 200:1644
55. Gianasi E, Wasil M, Evagarou EG, Keddle A, Wilson G, Duncan R (1999) Eur J Cancer 35:994
56. Howe-Grant ME, Lippard DJ (1980) Aqueous platinum (II) chemistry: Binding to Biological Molecules. Mercel Dekker, New York, 11:63
57. Yokoyma M, Okano T, Sakurai Y, Suwa S, Kataoka K (1996) J Contr Release 39:351
58. Nishiyama N, Yokoyma M, Aoyagi T, Okano T, Sakurai Y, Kataoka K (1999) Langmuir 15:377
59. Nishiyama N, Kato Y, Sugiyama Y, Kataoka K (2001) Pharm Res 18:1035
60. Nishiyama N, Okazaki S, Cabral H, Miyamoto M, Kato Y, Sugiyama Y, Nishio K, Matsumura Y, Kataoka K (2003) Cancer Res 63:8977

61. Bae Y, Fukushima S, Harada A, Kataoka K (2003) Angew Chem Int Ed 42:4640
62. Bae Y, Nishiyama N, Fukushima S, Koyama H, Matsumura Y, Kataoka K (2005) Bioconjugate Chem 16:122
63. Lee ES, Na K, Bae YH (2003) J Contr Release 91:103
64. Lehrman S (1999) Nature 401:517
65. Marshall E (1999) Science 286:2244
66. Cavazzana-Calvo M, Hacein-Bey S, Saint Basile S, Gross F, Yvon E, Nusbaum P, Selz F, Hue C, Certain S, Casanova J-L, Bousso P, Le Deist F, Fischer A (2002) Science 288:669
67. Aiuti A, Slavin S, Aker M, Ficara F, Deola S, Mortellaro A, Morecki S, Andolfi G, Tabucchi A, Carlucci F, Marinello E, Cattaneo F, Vai S, Servida P, Miniero R, Roncarolo MG, Bordignon C (2002) Science 296:2410
68. Hacein-Bey-Abina S, Von Kalle C, Schmidt M, McCormack MP, Wulffraat N, Leboulch P, Lim A, Osborne CS, Pawliuk R, Morillon E, Sorensen R, Forster A, Fraser A, Cohen JI, de Saint Basile G, Alexander I, Wintergerst U, Frebourg T, Aurias A, Stoppa-Lyonnet D, Romana S, Radford-Weiss I, Gross F, Valensi F, Delabesse E, Macintyre E, Sigaux F, Soulier J, Leiva LE, Wissler M, Prinz C, Rabbitts TH, Le Deist F, Fischer A, Cavazzana-Calvo M (2003) Science 302:415
69. Gebhart CL, Kabanov AV (2001) J Contr Release 73:401
70. Niidome T, Huang L (2002) Gene Ther 9:1647
71. Hassan Farhood H, Bottega R, Epand RM, Huang L (1992) Biochim Biophys Acta Biomembranes 1111:239
72. Wu GY, Wu CH (1998) J Biol Chem 263:14621
73. Boussif O, Lezoualc'h F, Zanta MA, Mergny MD, Scherman D, Demeneix B, Behr JP (1995) Proc Natl Acad Sci USA 92:7297
74. Behr J-P (1997) Chemia 51:34
75. Choksakulnimitr S, Masuda S, Tokuda H, Takakura Y (1995) J Contr Release 34:233
76. Fischer D, Bieber T, Li Y, Elsasser H-P, Kissel T (1999) Pharm Res 16:1273
77. Boeckle S, Gersdorff K, van der Piepen S, Culmsee C, Wagner E, Ogris M (2004) J Gene Med 6:2004
78. Osada K, Yamasaki Y, Katayose S, Kataoka K (2005) Angew Chem Int Ed 44:3544
79. Katayose S, Kataoka K, Ogata N, Kim SW, Feijen J, Okano T (eds) (1996) Advanced Biomaterials in Biomedical Engineering and Drug Delivery Systems. Springer, Tokyo, p 319–320
80. Katayose S, Kataoka K (1997) Bioconjugate Chem 8:702
81. Katayose S, Kataoka K (1998) J Pharm Sci 87:160
82. Wolfert MA, Schacht EH, Toncheva V, Ulbrich K, Nazarova O, Seymour LW (1996) Hum Gene Ther 7:2123
83. Dash PR, Toncheva V, Schacht EH, Seymour LW (1997) J Contr Release 48:269
84. Toncheva V, Wolfert MA, Oupicky D, Ulbrich K, Seymour LW, Schacht EH (1998) Biochem Biophys Acta 1380:354
85. Seymour LW, Kataoka K, Kabanov AV (1998) Self-assembling Complexes for Gene Delivery. Wiley, Chichester, p 219
86. Itaka K, Harada A, Nakamura K, Kawaguchi H, Kataoka K (2002) Biomacromolecules 3:841
87. Itaka K, Yamauchi K, Harada A, Nakamura K, Kawaguchi H, Kataoka K (2003) Biomaterials 24:4495
88. Harada-Shiba M, Yamauchi K, Harada A, Takamisawa I, Shimokado K, Kataoka K (2002) Gene Ther 9:407
89. Mishra S, Webster P, Davis ME (2004) Eur J Cell Biol 83:97

90. Merdan T, Kunath K, Petersen H, Bakowsky U, Voigt KH, Kopecek J, Kissel T (2005) Bioconjugate Chem 16:785
91. Kakizawa Y, Harada A, Kataoka K (1999) J Am Chem Soc 121:11247
92. Trubetskoy VS, Loomis A, Slattum PM, Hagstrom JE, Budker VG, Wolff JA (1999) Bioconjugate Chem 10:624
93. Carlisle RC, Etrych T, Briggs SS, Preece JA, Ulbrich K, Seymour LE (2004) J Gene Med 6:337
94. Meister A, Anderson ME (1983) Annu Rev Biochem 52:711
95. Kakizawa Y, Harada A, Kataoka K (2001) Biomacromolecules 2:491
96. Miyata K, Kakizawa Y, Nishiyama N, Harada A, Yamasaki Y, Koyama H, Kataoka K (2004) J Am Chem Soc 126:2355
97. Miyata K, Kakizawa Y, Nishiyama N, Yamasaki Y, Watanabe T, Kohara M, Kataoka K (2005) J Controlled Release 109:15
98. Erbacher P, Roche AC, Monsigny M, Midoux P (1996) Exp Cell Res 225:186
99. Fredericksen BL, Wei BL, Yao J, Luo T, Garcia JV (2002) J Virol 76:11440
100. Wang CY, Huang L (1984) Biochemistry 23:4409
101. Wagner E, Plank C, Zatloukal K, Cotton M, Birnstiel ML (1992) Proc Natl Acad Sci USA 89:7934
102. Plank C, Oberhauser B, Mechtler K, Koch C, Wagner E (1994) J Biol Chem 269:12918
103. Wagner E (1998) J Contr Release 53:155
104. Shoen P, Chonn A, Cullis PR, Wilschut J, Scherrer P (1999) Gene Ther 6:823
105. Li W, Nicol F, Szoka FC Jr (2004) Adv Drug Deliv 56:967
106. Farhood H, Serbina N, Huang L (1995) Biochim Biophys Acta 1235:289
107. Hui SW, Langner M, Zhao YL, Ross P, Hurley Chan K (1996) Biophys J 71:590
108. Harashima H, Shinohara Y, Kiwada H (2001) Eur J Pharm Sci 13:85
109. Kichler A, Leborgne C, Coeytaux E, Danos O (2001) J Gene Med 3:135
110. Suh J, Paik HJ, Hwang B (1994) Bioorg Chem 22:318
111. Zuidam NJ, Posthuma G, de Vries ET, Crommelin DJ, Hennink WE, Storm G (2000) J Drug Target 8:51
112. van de Wetering P, Moret EE, Schuurmans-Nieuwenbroek NM, van Steenbergen MJ, Hennink WE (1999) Bioconjugate Chem 10:589
113. Putnam D, Gentry CA, Pack DW, Langer R (2001) Proc Natl Acad Sci USA 98:1200
114. Kabanov AV, Bronich TK, Kabanov VA, Yu K, Eisenberg A (1996) Macromolecules 29:6797
115. Itaka K, Kanayama N, Nishiyama N, Jang W-D, Yamasaki Y, Nakamura K, Kawaguchi H, Kataoka K (2004) J Am Chem Soc 126:13612
116. Fukushima S, Miyata K, Nishiyama N, Kanayama N, Yamasaki Y, Kataoka K (2005) J Am Chem Soc 127:2810
117. Graham FL, van der Eb AJ (1997) Virology 52:456
118. Chen C, Okayama H (1987) Mol Cell Biol 7:2745
119. Tolou H (1993) Anal Biochem 215:156
120. Kakizawa Y, Kataoka K (2002) Langmuir 18:4539
121. Clapham DE (1995) Cell 80:259
122. Guyton AC (1992) Human Physiology and Mechanism of Disease, 5th ed. Saunders, Philadelphia, PA
123. Kakizawa Y, Furukawa S, Kataoka K (2004) J Contr Release 97:345
124. Kakizawa Y, Miyata K, Furukawa S, Kataoka K (2004) Adv Mater 16:699
125. Elbashir SM, Harborth J, Lendeckel W, Yalcin A, Weber K, Tuschl T (2001) Nature 411:494
126. Wu GY, Wu CH (1988) J Biol Chem 262:14621

127. Schatzlein AG (2003) J Biomed Biotechnol 2003:149
128. Wickham TJ (2003) Nat Med 9:135
129. Wagner E, Kircheis R, Walker GF (2004) Biomed Pharmacother 58:152
130. Nagasaki Y, Yasugi K, Yamamoto Y, Harada A, Kataoka K (2001) Biomacromolecules 2:1067
131. Yamamoto Y, Nagasaki Y, Kato M, Kataoka K (1999) Colloids Surf B Biointerfaces 16:135
132. Nah JW, Yu L, Han SO, Ahn CH, Kim SW (2002) J Contr Release 78:273
133. Torchilin VP, Lukyanov AN, Gao Z, Papahadjopoulos-Sternberg B (2003) Proc Natl Acad Sci USA 100:6039
134. Kataoka K, Harada A, Wakebayashi D, Nagasaki Y (1999) Macromolecules 32:6892
135. Wakebayashi D, Nishiyama N, Yamasaki Y, Itaka K, Kanayama N, Harada A, Nagasaki Y, Kataoka K (2004) J Contr Release 95:653
136. Oishi M, Nagasaki Y, Itaka K, Nishiyama N, Kataoka K (2005) J Am Chem Soc 127:1624

Author Index Volumes 201–202

Subject Index

Printing: Krips bv, Meppel
Binding: Stürtz, Würzburg